The
Odd Book
of
Data

TRANSLATIONS BY J. M. D. STEEN
ILLUSTRATIONS BY J. A. VAN DIJK

The *Odd* Book of of Data

by
R. HOUWINK
Wassenaar, The Netherlands

ELSEVIER PUBLISHING COMPANY
AMSTERDAM - LONDON - NEW YORK
1965

ELSEVIER PUBLISHING COMPANY
335 JAN VAN GALENSTRAAT, P. O. BOX 211
AMSTERDAM, THE NETHERLANDS

AMERICAN ELSEVIER PUBLISHING COMPANY, INC.
52 VANDERBILT AVENUE, NEW YORK, N. Y. 10017

ELSEVIER PUBLISHING COMPANY LIMITED
RIPPLESIDE COMMERCIAL ESTATE
BARKING, ESSEX

Library of Congress Catalog Card Number 65-13232

PRINTED IN U.S.A.

Introduction

EINSTEIN once said: 'That which is eternally incomprehensible to us in Nature is her comprehensibility'. In other words, however complex her laws may seem to us, Nature conducts her affairs according to simple principles, and all her creations – whether base matter or the stuff of life – are cast in a simple mould.

Whether we accept the general truth of EINSTEIN's maxim or not, we certainly encounter great difficulty in reducing Nature's multiplicity of forms to orderliness. For man – though her ablest observer – yet lacks the capacity to appreciate the sheer magnitude of number and extremes of dimension in which her parts disclose themselves.

We are usually able, be it with the aid of instruments, to set bounds to Nature's proliferation and to determine her chosen proportions, but... what do such computations mean to us? What impression, for example, can we form of the size of water molecules when we know that a thimble can contain billions upon billions of them? What are we to make of intra-atomic dimensions expressed in one hundred millionths of a micron, or of interstellar distances calculated in thousands of light years? When the scientist speaks of *nanoseconds*, do we – does even he! – pause to ponder that as many nanoseconds pass per second as seconds lapse in thirty years?

The logarithmic scale helps us to find our bearings, but does not bring us far along our course: it can hardly quicken the imagination. A more effective aid to the understanding of Nature's unimaginable dimensions suggests itself by way of comparisons or images which, in themselves, are more or less, but at least better, susceptible to the imagination. May this book – without striving after completeness – prove the soundness of the approach.

Whatever its shortcomings, the present treatment of the mutual numerical relationships of Nature's parts may perhaps claim some didactic merit, for its deliberate purpose is to simplify – and thereby render more memorable – those quantitative data which we, and particularly the young student of science, are often at pains to absorb and retain. It is surely no blemish, in this respect, that the choice of example often enters the realm of the comic.

Wassenaar, The Netherlands, 1965   THE COMPILER

Acknowledgements

In the first place I should like to express sincere thanks to the following people whose intensive and unstinting work has contributed toward the realization of this book, which touches on so many subjects.

Mr. J. L. BONEBAKKER, *Enschede*
Mr. K. L. VAN DEN BOS, *Enschede*
Professor H. ENGEL, *Amsterdam*
Dr. J. C. GERRITSEN, *The Hague*
Dr. E. F. J. JANETZKY, *The Hague*
Mr. J. JANSSEN, *Delft*
Professor C. W. KOSTEN, *Delft*
Mr. R. E. J. ZIECK, *The Hague*

Many more, too numerous to mention, have also played a role: without their help my job would have been considerably more difficult.

A special word of thanks is due to Mr. J. M. D. STEEN, *The Hague*, who did the translation. This was indeed more than a translation: Mr. STEEN turned out to have the talent to introduce a special and personal character into the texts, which perfectly fitted the sometimes extraordinary nature of the material. Should the *Odd Book of Data* be recognized as being more than the mere summing up of facts, then a great deal of the credit must go to him.

Many of the DATA collected in this ODD BOOK were originally found in published books and articles. It would, however, be misleading as well as impractical to mention all these sources, since the pertinent facts have often been combined, added to or adapted to meet the particular demands of the text. For these reasons, only the more general sources from which information has been culled are referred to below.

In some fields it has been difficult to make a selection from the large number of *noteworthy* facts and figures. In such cases, a *best choice* has been attempted on the advice of experts.

GENERAL SOURCES
  Encyclopaedia Americana
  Encyclopaedia Brittannica
  Der Grosse Brockhaus
  ENSIE Encyclopaedie, Amsterdam
  The Guinness Book of Records, London, 1962
  Larousse du XXme Siècle
  Scientific American
  Winkler Prins Encyclopaedie, Amsterdam
  The World Almanac and Book of Facts

ENERGY
  A. A. DE BOER, *Dissertation*, Leiden, 1962
  J. H. DE BOER, *Koninkl. Ned. Akad. Wetenschap.*, 1956
  P. PUTNAM, *Energy in the Future*, 1951
  U.N.O., *New Sources and Economic Development*, 1957

NUCLEAR FIELD
  Y. CHELET, *L'Energie Nucléaire*, Paris, 1963

UNIVERSE
  G. GAMOV, *La Création de l'Universe*, 1956

SOUND
  J. R. PIERCE, *Man's World of Sound*, New York, 1958

Contents

The Universe

Chaos and Creation

So far as astrophysicists can tell, by far the greater part of all *matter* in the Universe consists of free particles (protons, neutrons, deuterons, electrons) displaying no semblance of order. Only an infinitesimal fraction of one per cent appears to be organized in the form of atoms and molecules.

The conjugation of protons and neutrons into atomic nuclei was perhaps the most critical step toward the creation of discrete conglomerations of matter – from which the *celestial bodies* could take on their characteristic forms. According to one school of thought, this *essential* step could have been accomplished in less than a single hour, the electrons being added at a later stage.

A Yardstick for the Universe

Representing the Earth by one of the full stops (diameter 0.5 mm = 0.02 inch) on this page, our planet would be separated

– from the Moon: by the thickness of a finger (16 mm);
– from the Sun: by the length of a limousine (6 m);
– from the nearest star: by the length of the Rhine or the Ohio River (1500 km);
– from the Milky Way: by 200 times the circumference of the Earth (200 × 40000 km);
– from the Andromeda Nebula: by 4000 times the circumference of the Earth (the Andromeda Nebula is one of the Milky Way's 'nearest' neighbours; others are millions of times more distant!).

The Expanding Universe

Seen from the Earth, the galaxies are receding with speeds proportional to their distances. This can best be explained on the assumption that the Universe is continuously expanding.

The situation could be illustrated with the help of a rubber balloon with particles of sand adhering to its surface. On being inflated, the balloon would expand and the particles would move apart at a rate proportional to their increasing distance from the centre of the balloon.

The galaxies themselves are of almost inconceivable magnitude. Representing the Earth by a child's toy balloon, the milky way would still be about two million times larger than the Earth's actual size.

The Milky Way – the Escapist's Dream

Those wishing to 'get away from it all' will be attracted by the opportunities for privacy offerred by the Milky Way – the choice of at least thirty solar systems for each individual of the human race.

The occasional urge for companionship would present considerable problems, however. Inhabitants of the extreme limits of the galaxy, travelling by interstellar rocket at a speed of, say, one million km per hour would require a mere 100 million years to visit a 'next-door' neighbour... if he were at home!

Apropos:

> We, who once were wholly sure
> Escapism was immature,
> Now equate precocity
> With escape velocity.

Orbital Speeds

Travelling at the Earth's orbital speed about the Sun, a rocket would cover the distance between Paris and Rome (New York and Gander) in one minute.

At the Sun's orbital speed, this same distance would be covered in 6 seconds. The Moon, slow as it is, would need half an hour.

. . . a rocket would cover . . .

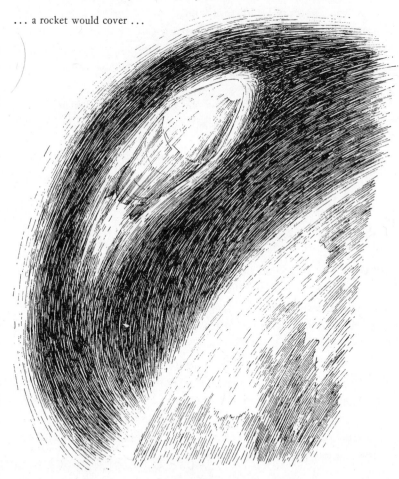

En Route

In travelling out to the Milky Way by light-velocity space-ship (300,000 km/sec), our escapist and his family could make a fleeting acquaintance with the sights of the cosmic inner harbour: our own solar system.

They pass the magnificent lighthouse, the Sun, in 8 minutes, the huge breakwater of Jupiter 45 minutes after departure and reach the unmistakable deap-sea buoy of Saturn with its circling reflectors in 90 minutes. It is five whole hours before Pluto looms up to mark the end of the roads and...the way to the stars in the boundless cosmic ocean.

Proxima (the nearest star, as the name suggests) surprises the travellers after four years of black emptiness, but this is after all but a short interval compared to the 150 years it will take before the surviving great-great- or, perhaps the great-great-great-grand-children will see the splendour of Rigel in Orion at close quarters.

The Tick of the Cosmic Clock

If the slow one-second tick of an old grandfather clock is given the value of one year, then the strokes of the pendulum which have actually measured out the seconds since Lincoln began his presidential term of office over a century ago, can be taken to represent the years which have passed from the Earth's creation to the present day.

By the same token, the clock would need to have been ticking away for at least 200000 years if it were now to 'tell the time' of the Sun's approximate period of existence, and would have to tick on for another 500 years to represent with every passing second a year of the Sun's anticipated residual capacity to support life on the Earth, *i.e.* to supply light and heat on the same scale as at present.

Planetary Years

Defined as a single revolution of a planet about the Sun, a 'year' differs from one planet to another.

Of three contemporary visitors to Earth from Pluto, Uranus and Mercury, each 25 'planet' years old:
– the Plutonian's name might have been written in hieroglyphics on a papyrus roll, having been born in the time of the early Pharaohs;
– his contemporary from Uranus might have been a playmate of HANNIBAL (200 B.C.);
– the Mercurian would just have celebrated his sixth birthday.

Sun-baked Cereal

Most of the Sun's rays originate in the *photosphere*, a 2000-km thick layer at the solar surface, which has a *grained* appearance when photographed from the Earth. The *grains* are thought to be the crests of gas fountains which spring from the interior of the Sun.

The Vatican astronomer SECCHI has compared the flaming surface of the Sun with a rice pudding, the individual grains of which can be as large as the Earth.

Looking Backwards in Time

When we observe a star which is, say, 5 light years distant, we see it as it was 5 years ago.

The light now reaching us from the star Pollux tells us about HITLER's rise to power in Germany and about the period of ROOSEVELT's New Deal, whilst a star in the Andromeda Nebula brings us, as it were, a 'visual greeting' from the period when *Homo sapiens* made his engravings in caves (15 000 B.C.).

The Earth

The Earth – Old, yet Fertile

Vain as womankind, the Earth
Tells not willingly her age;
But, unceasingly, the sage,
Armed with radiation gauge,
Calculates the half-life's worth,
And, retracing stage by stage
Rough steps of stony heritage,
Pries from prehistoric page
Her bleached certificate of birth.

Three times a thousand million years?
Conservative the estimate,
Say some! Yet 'tis commensurate
Ten times the human aggregate
When Christ was born. But there appears
No menopause. At constant rate
Of birth, mankind will emulate
The sum of Earth's rotation rate
About the Sun, as 1980 nears.

J. M. D. STEEN

The Terrestrial Tennisphere

To gain a rough idea of the relative size and extent of certain physical features of the Earth and its spatial environment, it is convenient to represent the globe by a tennis-ball. On this scale one finds that:

- the Earth's crust would be two and a half times the thickness of the wall of the ball;
- Mount Everest would appear as a barely visible bubble on the ball's surface;
- the great rivers Nile and Mississippi–Missouri would extend no further than half the length of one's little finger;
- the troposphere (where the weather phenomena occur) would be no thicker than a layer of water on the slightly moistened surface of the ball;
- the Moon might be represented by an eye, if the tennisball were placed at one's feet 2 metres (2.2 yards) away, though one's own body would shrink to the size of a virus;
- the Sun would be about the size of a large automobile 7 metres (8 yards) long, and it would be separated from the tennis-ball by about 700 metres (800 yards).

The Evolutionary Calender

If we imagine the entire history of the Earth to be condensed into the space of a calender year, ending in a few moments from now, then:

- seven months ago, in June, the first organic compounds were formed;
- at the beginning of October, the first signs of life might have been observed;

- in the second week of the present month, December, the verte-
brates appeared, and
- just over a week ago today, the mammals followed;
- man made his first entry on the scene a couple of hours before
midnight, and
- less than a minute ago, made his first entry in the book of recorded
history;
- as we hear the first stroke of midnight, COLUMBUS' discovery
of America is, perhaps, four seconds old.

The Earth's Solar Orbit

Jimmy, finding himself commanded to draw a diagram of the Earth's
orbit about the Sun, stepped to the blackboard and drew a circle.

'But Mr. Armistead, sir... you told us it was an ellipse!', exclaim-
ed little Mary.

'You are right, Mary, but as a matter of fact Jimmy's circle is a
tolerable representation of the Earth's elliptical orbit, its variations
from a truly circular path being contained, on his chosen scale, well
within the thickness of the chalk line.'

The World's Transport

If the entire mass of the Earth (6×10^{24} kg) were to be broken up
and transported by freight train, and if the 20-ton freight-cars of the
train passed an observer at the rate of one a second, the first load
would have had to have been checked through when the earliest
mammals inhabited the Earth (a hundred million years ago) if
the observer were now to be able to witness the passing of the
last car.

Stardust

Every day the Earth's surface receives some 15000 tons of meteorites and other solid matter from outer space – sufficient to fill about 500 large railway boxcars. But for the effects of erosion from wind and rain, and river drainage, this constant deluge of 'dry rain', which has probably continued since the Earth's creation, would have produced a continental dust-layer 20 metres (60 feet) deep.

Though the daily shower includes an estimated 20 million visible particles larger than a pinhead, the danger of massive bodies penetrating to the Earth's surface without being consumed by the heat of atmospheric friction is fortunately very small. Nevertheless, meteorites of 60 and 150 tons, such as have been unearthed in Greenland and South Africa respectively, and a colossus of the kind which dug a 1200-metre (1300-yard) crater in the soil of Arizona, U.S.A., represent an occasional hazard.

World Awash

Roughly two-thirds of the surface of the Earth is covered by water. The area of the Pacific Ocean alone is actually 25% larger than that of all the land surfaces of the world taken together.

A reason why the land and the sea level approach each other more and more is the washing of land into the sea by rivers.

Taking the average height of the USA above sea level as 800 metres (2600 ft) and putting the rate of erosion at 6.3 cm (2.5 inches) in 1000 years, the volume of USA above sea level will be washed to the ocean within 12 million years: this is only something like one tenth of the period since which mammals populate our planet.

Our Rain Supply

All the water present in the Earth's atmosphere (estimated at 13 million tons) could be stored in the basins of Lakes Huron and Michigan.

Evenly distributed over the surface of the Earth, it would form a layer of 2.5 cm (1 inch) thick. It has been calculated that this quantity of water actually falls on the Earth in about eleven days: consequently this is the average life cycle of a water molecule in the air.

European Decline

If Western Europe continues to subside at its present rate of about 2.5 cm (1 inch) every ten years, the top of the Eiffel Tower will become a small island in the Atlantic in roughly 120000 years time.

Brine Mine

If all the salt dissolved in the seas of the world were crystallized, a layer 45 metres (150 feet) thick could be spread out over the entire land surface of the globe. This volume of salt would be fifteen times greater than the European land mass above sea level.

Coastlines

Though the Italian mainland is three times larger than the area of Greece, the relatively unaccidented coastline of Italy is about 20% shorter than the highly irregular littoral of her Hellenic neighbour.

Physics

What Einstein's Relativity Theory Tells Us

According to EINSTEIN *length*, *mass* and *time* are dependent on the velocity of motion, relative to the observer.

Length An observer 'sees' the length of a moving body shorter in the direction of motion, the contraction being a function of its speed:
– in the case of an automobile (100 km per hour), the observer would not be conscious of the contraction, which corresponds to no more than the diameter of an atomic nucleus;
– the same is true of a jet plane (1000 km per hour), whose length should appear one atom smaller than it actually is;
– the apparent contraction of an interplanetary rocket travelling at 400000 km per hour would amount to 0.01 mm;
– to an observer on another planet in the solar system the Earth would appear 6 cm shorter in the East–West direction, due to its rotational speed of 30 km per sec about the sun;
– if the Earth were to rotate 10000 times faster about the sun, a six-foot (1.8-metre) man, lying in a bed placed parallel with the Equator, would appear to the observer to be only three feet (nearly one meter) tall. If his bed were turned at right angles to the Equator, he would seem to regain his original length, but the breadth of his body would then appear to be halved. These observations would be made without any stresses being involved.

Mass Mass increases with speed, becoming infinite at the velocity of light. At extremely high speed (say 10^{-14} percent below the velocity of light), a revolver bullet would have the mass of a fully laden freight car.

Time To an outside observer, the passage of time is reduced by velocity. The return journey to a star, nine light years distant, by a rocket travelling with a speed of 10^{-8} percent below the velocity of light, would take slightly more than 18 years. The heart and lungs of the rocket's crew, and the clock they carried with them, would appear, however, to an outside observer to work so slowly, that the 18 years would seem to last but a few hours. The observer would expect the rocket to have completed its two-way journey between, say, lunch and dinner on the same day, but at the same time he would observe that the families of the rocket's crew would have consumed lunch and dinner well over 6500 times.

Non-Euclidean Space

According to relativity theory space is *curved* (non-Euclidean) in the vicinity of massive bodies. In such space the ratio between the circumference and the radius of a circle deviates from 2π.

If, after accurately measuring the circumference and radius of a large circle drawn on an open space, like Piccadilly Circus or Times Square and thence determining π experimentally, a small insect were to crawl into the circle, the mass of this creature would impart sufficient curvature to the surrounding space to alter, say, the thirtyfifth decimal place of π. The 765 remaining decimals, as calculated by a Danish schoolmaster (see p.76) would then of course become meaningless.

Deflection of Light

When the light from a star reaches the Earth after passing through the immediate atmosphere of the Sun, the angle of deviation induced in the path of the rays by the Sun's gravitational attraction approximates to the angle subtended by a 2-metre (6½-foot) pole located 200 km (125 miles) from the point of reference.

The One-Second Sprint

An air of expectancy hung over Trafalgar Square. A huge crowd of Londoners sized up the eight competitors for a novel international event: 'The One-Second Sprint'.

The first participant, *Herr Ottokar*, fresh from triumphs on the Nürburgring, flashed into the Square and steered four screaming tyres through the starting gate at a brisk 250 km per hour (150 miles per hour), finding himself a second later 70 metres (76 yards) farther on.

M. Bruit, true to his gallic temperament, made a great deal of commotion, but penetrated a mere 340 metres (1100 feet) into the Strand, after breaking his own sound barrier.

... flashed into the Square and steered four screaming tyres through the starting gate ...

The following participant, *Mr. Brown*, mimicking the molecular motions he had himself discovered, contrived within his allotted second to pass the gates of the Tower, fully 2 km (1.3 miles) from the start.

Next, the lady-representative of the venerable firm of *Terra* (long-standing consultants in daily and yearly orbiting) reeled into the Square. Developing the speed of the daily rotation about her axis, she executed a prodigious bound which doubled the distance recorded by her forerunner.

Miss Ile summoned her accumulated experience of hasty escapes, and managed a splendid burst to Croydon, 12 km (7.5 miles) away.

It was left to the diminutive *Alec Tron*, however, to dispose of Miss Ile's challenge. Whirling himself about an obliging atomic nucleus he flew off at a rare pace, arriving over St. Peter's in Rome just as 'time' was called... a mere 2000 km (1200 miles), if you please!

Amid the popular enthusiasm, the announcement of the next competitor was barely audible, but a knowing smile crept over one or two faces, although the external appearance of the next competitor gave little hint of the sensational climax in store for the onlookers.

Fully two thousand times fatter than the front-running Alec Tron, *'Quick' Neutron* was naturally regarded as a comic turn to round off the proceedings until... he vanished into thin air! The blank astonishment of all in the Square can be imagined when, shortly after his departure, a call from San Francisco announced that 'Quick' had arrived over the Golden Gate Bridge precisely one second after his disappearance. Incredulous timekeepers calculated his trajectory: a fantastic 10000 km (6000 miles), which meant that 'Quick' Neutron had outpaced the flamboyant Herr Ottokar by more than 140000 times!

With an evil glint in his grecian eye, *Dr. Gamma* sped quick as a flash from the starting gate, and was lost to view; only to re-appear a little less than a second later at his point of departure, claiming that he had in the meantime finished seven round the world trips. 'Failed to qualify!' came the verdict. Indignant at being taken for an illusionist, the sinister Greek fell to muttering, threatening to irradiate the judges with a dose of his own medicine, but neither the Committee nor the public had an ear for so unsporting a competitor.

A Fly on Board

Whenever a fly alights on an ocean liner of about 35 000 tons, the ship tends to sink lower in the water by one tenth the thickness of an atom (0.1 Å) – this can be measured at present by means of an electrostatic capacity meter. If the fly lands on the handrail, say 15 metres (17 yards) from the centre-line of the ship, the resulting downward deflection of the ship on the same side will be about twenty times greater (unless the vessel is efficiently stabilized). In fact, it is not even necessary for the fly to touch the ship at all. If it merely hovers just above the deck, the vertical pressure of the airstream generated by its wings will have practically the same effect on the ship.

Whenever a fly alights
on an ocean liner . . .

One Second of Arc

One second of arc (one sixtieth of a minute) is the angle which would be formed by two radians of 200 km (125 miles) subtending a pole of 2 metres (2.2 yards) length.

Free Fall

A helicopter pilot who thought to take the easy way up Mount Everest (29 000 feet = 8880 metres) but mistook the characteristic wind-blown plume of snow at the summit for the firm foothold of the pinnacle, would have threequarters of a minute to reflect on his slight error of judgement as he plummeted down to sea level.

Advancing the Clock

The watch of a visitor to the top (102nd floor) of New York's Empire State Building (380 metres = 1250 feet above street level) tends to gain as a result of the decreased braking effect of the Earth's magnetic field. The increased tempo of the clockwork is, however, not too significant (about $5 \times 10^{-4}\%$): 70 million years would have to elapse before the watch had gained a full second.

Atoms and Molecules

Nature's Building Blocks

Ninety stable atoms in various combinations are the permanent building blocks of molecules, of which at present about a million different kinds are known. This is but a very modest number when one considers the seemingly endless possibilities of combination in

... each human being possessing his own 'tailormade' set of macromolecules.

the length, arrangement, shape and composition of macromolecules. So rich is the potential of protein molecules alone, that one can readily imagine each human being possessing his own 'tailor-made' set of macromolecules.

The flora and fauna of the world are for a great part composed of macro-molecules, so that it is not surprising that we meet such variety in their form and appearance. Nevertheless, it is certain that the 200 000 different kinds of plant and the million distinct sorts of animal which are so far known to have evolved, represent the merest fraction of the possibilities of differentiation.

Forces of Attraction in Matter

2–3 grammes of protons (the weight of a dime), placed on each of the opposite poles of the Earth would repel each other with a force of about 30 tons. There must thus exist within the atoms very great forces of attraction to hold nuclei together since they would otherwise explode.

How Small Molecules Are

Most molecules are of such small dimensions* that even scientists experience considerable difficulty in visualizing their relationship to the every-day physical world. For this reason, a few *data* will be devoted to this problem by considering molecules in line, on surfaces, as individual units, and in bulk.

A Chain of Iron Atoms A chain composed of as many iron atoms as there are people in the U.S.A. (180 million) would extend for a distance of approximately 2 cm (less than 1 inch).

* The number of molecules in 1 cm³ of water is about 3.3×10^{22}.

A Monolayer of Molecules If every square metre of the Earth's dry surface were occupied by six men, their total would approximate the number of molecules in a monolayer of air one centimetre square (1 000 000 000 million).

A Teaspoonful of Molecules One teaspoonful of water contains at least as many molecules as the Atlantic Ocean contains teaspoonsful of water.

Columbus's Water Molecules If COLUMBUS had emptied a glass of water in the ocean, and assuming that this water were now thoroughly mixed and distributed through the seas of the world, then each glass of water taken from the nearest tap or other source of water would now contain up to 250 molecules of the original contents of COLUMBUS's glass.

A Handful of Snow If all the molecules in a handful of snow were magnified to the size of a pea, there would be snow enough to blanket the entire surface of the Earth to a thickness covering the Eiffel Tower or the Rockefeller Center in New York, *i.e.* 300 metres (330 yards).

Hospitality with Molecules Peter looked askance at John who was just polishing off his umpteenth glass of whisky:
  'Steady on, old man... leave a drop for the stragglers!'
  'Shteady on?' echoed John, 'Why, there'sh heapsh of the shtuff... hic. Tell you what, shport, I'll dish out a thoushand moleculesh per shecond to every living shoul for the nexsht thirty odd yearsh... hic.'
  John upturned his empty glass and shook a single drop onto Peter's palm. 'Take care of the dishtribution, shport!'

Counting Molecules Counting by hand the individual molecules composing 1 cm³ of water at the rate of one per second, the whole of recorded history would still represent the merest fraction of the time necessary to carry out the operation.

If the task had been finally completed today by an army of counters equal to the population of Toulouse (France) or Toledo (Ohio), the 300000 participants would have had to have begun counting the water molecules three thousand million years ago, *i.e.* at about the time the Earth is now thought to have been created.

The Motion of Electrons

Electrons travel along their orbital paths with a velocity of 2000 km per second, *i.e.* at a speed about one hundred times greater than that attained by any rocket or missile yet launched by man.

In a conductor, however, their progress may be very slow indeed. For example, when a current of 1 Ampère flows through a copper wire with a cross-sectional area of 1 mm², the speed of the electrons is as if a walking man would cover 6 metres (6.6 yards) per 24 hours. Their forward motion is then like that of a very viscous fluid pressed through a narrow-gauge tube; it is only the tremendous number of electrons in transit which results in the transfer of an appreciable amount of energy.

Consider now a 100-Watt lamp burning for one second at 100 volts. If the displaced electrons were to be represented by an army of soldiers marching shoulder to shoulder in ranks half a metre apart, each rank would contain the population of the Benelux countries (22 million) and the front and rear ranks of the column would be separated by the distance between the Earth and the Sun (about 150 million km or 93 million miles)!

The Size of the Electron

Inconceivably small though individual molecules may appear to us (see above), they are still of vast proportions when compared to electrons.

If, after the Flood*, NOAH had set himself to string electrons on a thread at the rate of one a second for eight hours a day, the chain so formed would today still be only two tenths of a millimetre long! Composed of hydrogen atoms, the chain would be something like 25 metres (80 feet) by now.

* Supposed to have taken place around 3000 B.C.

The Empty Atoms

If one could enlarge the nucleus of a carbon atom to the size of one of the airports of Paris (or Idlewilde, New York), the electrons in the outer shell of the atom could be thought of as small sports planes circling around over Oslo and Algiers (Gander and the Bermuda Isles).

If one could compress *empty* carbon atoms until the electrons touched the nuclei, one would obtain a material of such extremely high density that a coat-button made of it would be comparable in weight to four fully laden ships of the size of the *Queen Elizabeth* (83 000 BR tons).

Hitting the Uranium Nucleus

In the process of uranium fission, the chance of a single quick neutron hitting and splitting a uranium nucleus is so small that the average distance travelled by such neutrons before this effect is achieved is relatively very great indeed.

If we represent the diameter of the uranium atom by the ground-level cross-section of a giant California redwood tree (10 metres = 11 yards), the uranium nucleus will resemble a mere grain of sugar (1 mm across), and the quick neutrons will be 7/100 as small as this again. Consequently, most of these neutrons pass through a vast forest of redwoods (uranium atoms) before striking a grain of sugar (uranium nucleus). In fact, on our chosen scale, the average distance travelled by quick neutrons before collision with uranium nuclei would be like a mighty forest of (closely packed) redwoods, covering twenty times the distance between the Earth and the Moon (*i.e.* 20×350000 km).

Giant Molecules

If the nucleus of a carbon atom were to be represented by an apple pip which was placed in the centre of a large stadium, the electrons on the outer shell would occupy orbits as far from the centre as the spectators in the rearmost seats.

A molecule of rubber incorporating carbon atoms on the same scale would stretch the distance between London and Athens (New York and Gander).

Botany

Flower Power

Each year approximately forty times more carbon is used in photo-synthesis of carbon compounds from carbon dioxide by plants than is extracted by man from the world's coal mines.

Every Man's Own Field

Despite the fact that two-thirds of the Earth's surface is covered by water, there are still 5 hectares (12½ acres) of land to every inhabitant of the world. However, of this individual quota, one hectare is too cold for cultivation, one too mountainous, one too barren, and one too lacking in water. Consequently, there remains but one hectare per person for the production of food... but at present only half this area is actually being exploited.

The sea supplements the yield of this half hectare by providing the food-equivalent of one quarter the area of a tennis court.

Grain Enough

The amount of grain produced surplus to domestic and trade re-quirements in certain areas of the world (principally North America) is currently running at 130 million tons per year. This surplus alone could provide calories enough to support the entire population of the world for two months.

The annual increase of this surplus could satisfy the nutritional requirements of 40 million persons for twelve months, *i.e.* 85% of the present annual increase in the world's population.

Agriculture Here and There

American grain farmers are estimated to produce individually an average of 100 kg (220 lbs) of wheat per hour. The same rate of production currently occupies 17 farmers in Chile, 24 in Colombia or Pakistan, and 50 in Japan.

It is therefore not surprising that in the underdeveloped countries 60–80% of the population is engaged in agriculture, whereas in the United States 10–20% of the labour force not only satisfies national requirements, but provides huge quantities of agricultural produce for export, including 10 million tons of grain surplus to domestic needs.

Areas Involved in Cereals Production

The area of the United States exploited for the cultivation of cereals (wheat, rye, barley, oats, maize, millet, sorghum and rice) is nearly as large as the surface area of France and Italy combined.

The area under cereal cultivation in Italy corresponds in turn to the total area of the Benelux countries (Belgium, The Netherlands and Luxemburg). Luxemburg itself represents about half the area given over to the production of cereals in Belgium and The Netherlands.

The Farmer's Trailer

If the farmers of the United States and Western Europe brought their produce to market once a year, the American farmer would have an average load of ten tons, whilst the European would transport about three-quarters of this weight.

The American's trailer would contain an average 6.6 tons of cereals and a half a ton of potatoes. The European would be carrying 0.9 and 2.7 tons of these commodities respectively.

Both farmers would market about a half a ton of meat, but the European would offer nearly 10% more milk for sale than the American (2.3:2.1 tons).

Milky Weigh

In countries like Denmark and Holland where dairy farming is carried on very intensively, the average annual yield of milk per cow is roughly ten times the live weight of the animal (about 4000 litres = 840 gallons). In India, the annual yield per cow averages only about twice its body weight, but if account is taken of the extreme leanness of the Indian milch-cow, the comparable equivalent yield is only one-tenth of that of the best European stock.

Annual Trek

In the countries of the European Common Market (Belgium, France, Germany, Italy, Luxemburg and The Netherlands) as many farmers leave the land each year to take up other occupations as there are farmers in The Netherlands (150000).

Clockwise

Botanists have found that many flowering plants yield their nectar at certain fixed hours of the day – the timetable is unerringly adhered to by the bees. Wild mustard and certain kinds of dandelion receive a host of callers at about 9 a.m. Traffic to the blue cornflower is heaviest around 11 a.m., whilst for the red clover 1 p.m. is the preferred visiting hour, and for viper's bugloss, 3 p.m.

History Enshrined in Wood

Schematic cross-section of a California Redwood, showing how the annual rings provide us with a historical index.

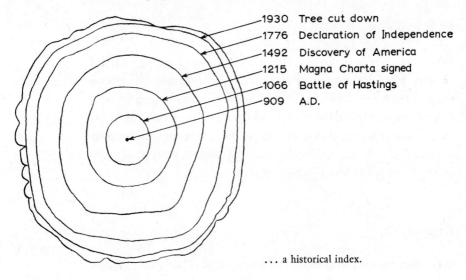

1930   Tree cut down
1776   Declaration of Independence
1492   Discovery of America
1215   Magna Charta signed
1066   Battle of Hastings
909   A.D.

... a historical index.

Biology

Small Beginnings

The number of female egg cells necessary for the next generation of the human race could be contained within a hen's eggshell. The corresponding quantity of male sperm would occupy a volume no greater than that of a pepper grain.

Is there here perhaps a possibility for mankind to secure its continuity in the event of a large-scale nuclear war?

Birth Control Needed

A cholera bacterium divides every thirty minutes to produce two complete entities. This form and rate of reproduction means that, in the space of only 24 hours, each cholera bacterium can be the progenitor of 10 000 times more of its kind than the total human population of the world.

The offspring of the four-inch-long Giant African Snail is so prolific that, if the fully grown progeny of a three-year period of reproduction were arranged in a straight line, the first- and last-born would be separated by a distance equivalent to two million round trips from the Earth to the Moon (over two light months). If the young were born even at the rate of one every second, the first of such a progeny would be 500 million years older than the last.

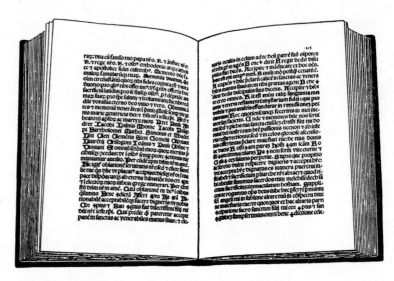

... *i.e.* roughly 500 times the letter count of the Bible.

The Library of Heredity

The submicroscopic bearers of hereditary characteristics, which we call genes, are long-chain molecules. They may be thought of as scrolls upon which the genetic language is written, the four letters of whose alphabet correspond to atomic groups of four different kinds.

A particular sequence of four letters on a portion of the molecular chain constitutes, as it were, a *file* in the *library* of heredity, and controls one or more characteristics, whilst each complete chain can be faithfully reproduced and passed on to a succeeding generation.

This hereditary *filing system* becomes progressively more voluminous as one passes from the lower to the higher forms of life.

In a small virus the *scroll* is of the order of o.003 mm in length, and contains about as many letters as half a page of newsprint (8000).

In the T_2-bacteriophage the information capacity of a 300-page paper-back novel (say, 500000 letters) is packed onto 0.1 mm of molecular chain.

In the mammal the uncoiled chain would span a full metre, and is able to accommodate some 3000 million letters, *i.e.* roughly 500 times the letter count of the Bible.

An interesting feature is that in the case of the virus it has been shown that changing one of the letters of its genetic language by nitric acid–corresponding to making one printing error in its hereditary file–would make this file unreadable.

Suspended Animation

A remarkable example of *suspended animation* is provided by certain bacteria found embedded in the dried salts of mineral-water springs at Bad Nauheim in Germany. After a period of dormancy estimated at no less than 300 million years, these bacteria were not only readily revived by moistening, but immediately proceeded to multiply!

Bacteria by the Hundredweight

Agronomists have calculated that the weight of bacteria in the top 30 cm (12 inch) of soil of a fertile meadow is roughly equivalent to the weight of the cattle that can be adequately supported on it. The bacteria content of soil can thus serve as an index of productivity.

So dense is the concentration of bacteria in the top layer of fertile soil that one single saltspoonful (say, 1 cm³) of it may often contain a population surpassing that of the human race (upwards of 3000 millions, living at present on all the soil available.

High Performance

Of all living creatures the hummingbird has the fastest rate of metabolism – actually a hundred times faster than that of some large animals. At its high cruising speed of 55 km per hour (35 miles per hour), the energy-output-to-weight ratio compares favourably with the performance of a modern helicopter flying at the same speed.

To support its very active mode of existence the hummingbird must convert relatively huge quantities of plant sugar to energy. Even so, the tiny creature would rapidly *burn itself up* if it did not go into *hibernation* during the hours of darkness, in which period its metabolic rate drops to a fifteenth of the daytime value.

Vivace ma non Troppo

If the metabolic cycle in the body cells of the elephant were to be accelerated to the same rate as in the tissues of the mouse, this largest of land mammals would be unable to effectively dissipate the heat developed by the oxidation of its stored fats and sugars. It would die in a matter of minutes from the effects of overheating.

Thin Red Line

In a normal human adult male, one cubic millimetre of blood contains about five million red corpuscles and between five and ten thousand white corpuscles. If all such cellular bodies from the total blood supply of one man were arranged in a single continuous thread, it would extend approximately seven times around the Earth.

Harmless Bleeding

So richly populated with red corpuscles is our bloodstream, that we can continuously lose large numbers of them without the slightest ill effect. Even if the rate of loss were one a second, it would take a million years or so to deplete the circulation of such corpuscles – and then only if the natural process of replenishment were inactivated.

Indeed, at this same rate of loss, a man would have to 'bleed' since C A E S A R came to power (two thousand years ago), before he had lost anything like as many red corpuscles as are normally withdrawn from a blood donor at one session.

Fish's-Eye View

When the sun rises just above or is just about to set below the horizon of the sea, its rays are bent to 49° from the vertical as they pass through the surface of the ocean. A fish's view of the whole scene above the surface of the sea is thus compressed within a cone of only 98°.

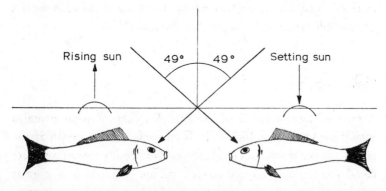

A fish's view compressed within a cone of only 98°.

Earning her Keep

In collecting one pound of honey, a bee may well fly a distance roughly equivalent to going twice round about the Earth.

Vital Cycles

Half of the water in the human body is renewed within two weeks. The turnover of red blood cells is even more rapid, half of them being replaced in seven days.

Comparing Heads

The gigantic head of the statue of Christ which commands the harbour of Rio de Janeiro weighs about thirty tons.

The head of the submicroscopic bacteriophage is probably the smallest distinguishable object in Nature which can justifiably be so called – so small, indeed, that any idea we might have about the size of our own heads lying roughly midway between these two extremes would be hopelessly astray. Even a pinhead represents a volume fully 1000 times larger than the mean!

Slow Death

A piece of human tissue of the size of a cube of sugar contains something like 250 million cells. If our cells were to perish at the rate of one per second, total extinction of our bodily tissue (complete metabolism) would require about a million years. As it is, a great number of our cells are being broken down at any given moment.

Honey Free of Charge

An entomologist has calculated that there are bees enough in the United States for each American citizen to maintain a hive of half a million of them.

... to maintain a hive of half a million of them.

... appeals to Mr. X's enemies as a very efficient weapon ...

Murder by Radiation

The lethal radiation dose for Mr. X, who weighs 70 kg (155 lbs), is 700 rads, which means an energy input of five thousand million ergs (120 calories). Consequently, radiation appeals to Mr. X's enemies as a very efficient weapon, the energy required to effect his demise being merely equivalent to raising their victim's body temperature by 0.002° C or causing him to jump from a table 1 metre (3 feet) high.

Notwithstanding this advantage of their chosen weapon, however, Mr. X's would-be assassins decide to abandon it when they hear that every living person would have to 'fire' one million quanta of the most concentrated kind (γ-radiation of 1 MeV) at Mr. X before he would be certain to succumb.

Fortunately, the Earth's natural level of radiation is normally low; each individual receives 'only' about 20000 quanta per second, at which rate it would take over 5000 years to kill a man in this natural way.

Protection against Radiation

We can protect ourselves against radiation of different kinds simply by shielding ourselves with paper of appropriate thickness:

– against α-particles (helium nuclei) a sheet of cigarette paper would suffice;
– against β-particles (electrons) a couple of sheets of ordinary note paper would be adequate;
– against quick neutrons a book like the present one would be protection enough;
– against γ-rays (electromagnetic waves) we should need a thick volume, say a 1000-page Bible.

Mankind

Eden and After

The entire progeny of ADAM and EVE (by which is meant the sum total of all human beings living and dead) is estimated to be only twenty-five times larger than the present human population of the world, say, 3300 million.

Host to the World

To accommodate all the present inhabitants of the world at a communal banquet, a row of chairs would be needed equal in length to fifty times the circumference of the Earth. It would be a restless

To accommodate all the people of the world at a communal banquet . . .

company, with diners constantly departing (to their graves), and others arriving (from the cradle) to occupy the vacant seats. The number of newcomers would, however, dominate to such an extent that extra seating space would have to be provided at the rate of 4 kilometres (about 2½ miles) per hour, just permitting the host to welcome each new guest at the table with a brief (1½ seconds) handshake.

By the year 2000, twice the original number of chairs would be needed, and the host would have to equip himself with a bicycle to greet new arrivals, who now take their seats as fast as his four-per-second mechanical chair-dispenser can set them down!

Storing-up Mankind

The entire world population could be stored up in Lake Windermere (Lake Candlewood, Connecticut) or – if you prefer – be packed shoulder to shoulder on the Isle of Wight.

If this mass of Humanity could be so condensed that no space remained between the nuclei and the electrons composing each individual, the 'pure' matter so formed could be contained in a single matchbox: the fantastic weight of its contents would promptly bore through the earth's crust.

The Gambling British

In 1962 the British public gambled on the outcome of sporting and other events to the extent of £15 ($40) per head of population, which worked out to well over £850 million ($2335 million) in total. This prodigious total of stake money, 65% of which was wagered on horse-racing and 10% on football, is about twice the amount annually devoted by the United States to foreign-aid programmes.

Life Expectancy

There seems to be a significant correlation between the general level of industrial development in a given area of the world and the average life expectancy of its inhabitants. In the highly developed countries of North America and Western Europe, both sexes may now expect to live an average 70 years – almost twice as long as the mean lifespan of the populations of economically depressed countries like India and the Congo Republic.

Where considerable progress towards modern exploitation of natural resources has been made, the mortality rate shows a clear tendency to decline. In Brazil, for instance, the average life expectancy has now risen to 52 years – roughly intermediate between the examples mentioned above.

The Beggar's Opera

U.S. Secretary of State, DEAN RUSK, has observed that in the General Assembly of the United Nations Organization, certain countries, together representing less than 10% of the world's population, and contributing less than 5% to the Organization's operating costs, are able to command a two-thirds majority of the votes.

Hungry

Whilst annual surpluses of one or another kind of agricultural produce are rather the rule than the exception in North America and Western Europe, 30% of the population in other areas of the world are permanently underfed and 10% are currently on the verge of starvation.

... the French Revolution ...

Can We Avoid a Second French Revolution?

Just prior to the French Revolution, the 25 million inhabitants of France were financially and economically dependent on about 250000 nobles and clericals, *i.e.* effective power was in the hands of only 1% of the population. This was capitalism in an extreme form.

At present 85% of the world's capital is controlled by roughly 20% of its population, but due to the disproportionate rise in the birthrate of the 'have-nots' it is estimated that by 1970 as much as 90% of this capital will be concentrated in the hands of only 8% of the population.

Spencer's Thirteen Spades

'You'll never believe this!' With unconcealed relish Spencer laid his contract hand, card for card, on the table... all thirteen spades! 'I take it you'll concede my *slam*, gentlemen?' he asked breezily of his opponents, and then addressed the senior member for the opposition: 'Your deal next, Stuart, make it the same again!'

Dr. Stuart wouldn't have been a statistician, however, if he had neglected an opportunity to discourse on his pet subject – probability.

'I should be happy to oblige, my dear Spencer, but I can offer you little hope of an immediate repetition of your abominable luck. Even if you played forty hands of bridge every evening, you would need – mathematically speaking – to have commenced playing with the primitive types of prehistoric mammal (eleven million years ago), since Homo sapiens had yet to arrive on the scene by then! Then you would have had a statistically justified expectation of personally drawing all thirteen cards of *any* suit.'

'But if you are going to insist on *spades*, Spencer, I'd better choose another example to bring home to you the whole sombre truth. If every living person in the world (3300 million) participated in a marathon drive of two hundred hands, only one of this giant company of odd players could be reasonably expected to draw a complete suit of cards. Not until the marathon had been played four times over would there have been a statistical likelihood of any complete suit drawn being *spades* rather than one of the other three suits. And then... if it is precisely you, dear Spencer, who is to have a reasonable expectation of finding all the spades in your hand, you must be prepared to attend over 3000 million such marathons! Chin up, Spencer – here comes the first card!'

A Look at Monte Carlo

The longest consecutive run of even numbers in the history of roulette at Monte Carlo is a series of twenty-eight. The mathematical odds against a series of twenty-nine even numbers occurring are so great, that the wheel at Monte Carlo would have had to have been spun since prehistoric times (about 6500 years ago) to give a statistically justified expectation of such a sequence occurring once.

The Americans in the One-Town World

In a community of 1000 persons in which the various nationalities and creeds were proportionally represented, the main language would be Chinese, spoken by 170 persons, and the next in importance English, used by 100 citizens, of whom 60 would be American. Only 390 members of this microcosmos would be classified as Christians, whilst 80 would be practising communists and 370 communist-dominated.

The 60 Americans would have an average life expectancy of 70 years as against 40 years for the remaining 940 members. The Americans would receive half the income of the entire community and would individually own possessions having fifteen times the value of the total property of the non-American members.

More Letters

It is estimated that by 1982 the world's postal services will have to cope with twice the number of letters now handled.

The 'undercrowded' U.S.A.

If all the inhabitants of the world were concentrated in the U.S.A., the resulting density of population would still not exceed that of The Netherlands or Belgium (300 per km²). Nevertheless, overcrowded though the Low Countries can now be said to be, the opposite is hardly true of the United States, since her present density of population is roughly equal to the world average (20 per km²).

... present density of population is roughly equal to the world average ...

Urbanization

By the year 2000, when the present population of the world will have doubled, only some 10% of the human race will remain employed on the land.

In a highly developed country like the U.S.A. (and perhaps at that time also Europe), every thousand new inhabitants will require eight schools, one hospital bed and two police officers. For every extra car added to the regular traffic of a city, two trees will need to be planted to compensate for the additional air pollution.

Manufactured Man

In an experiment conducted within the framework of the U.S. space development programme, seventeen volunteers lived entirely on synthetic food for a period of several months. At the conclusion of the experiment, all were found to be in excellent physical condition. The conclusion drawn from this and other researches is that one synthetic-food factory of moderate size could adequately feed the population of a large city.

Filling the Cavities

According to researches conducted by the World Health Organization, adequate control of oral hygiene requires the services of one dentist to every thousand persons. To realize this objective on a world-wide scale would entail the provision of as many dentists as there are citizens registered in Paris or San Francisco (approaching 3 million).

Inspired Genius

At every breath, each of us is likely to inhale some 50 million mole-cules of air* exhaled at some time by LEONARDO DA VINCI or, for that matter, by any other person who lived for 65 years and whose breath has been well distributed throughout the atmosphere.

* Oxygen, nitrogen, carbon dioxide, etc.

High-Pressure Heels

A woman of average size supporting her weight on one stiletto-heeled shoe, exerts a local pressure on the ground similar to that act-ing on the walls of an ordinary high-pressure steam boiler (75 atm). In the case of a heavily built representative of the fair sex, weighing about 100 kg (220 lbs), the pressure exerted by the tip of her heel would approach that attained in a modern heavy-duty boiler (140 atm).

Metropolis Before Christ

Until about 100 A.D., the largest city the world had known was Knossos on the Mediterranean island of Crete, thought to have been founded in the 15th century B.C.; nowadays, Knossos would com-pare with a small provincial town with some 80000 inhabitants (Orleans, France; Compton, California).

Quick on the Draw

World production of cigarettes is currently quite sufficient to pro-vide every living person with two cigarettes a day (2400000 million per year). In the United States industry 'rolls out' six cigarettes per head of population per day.

The Flowing Bowl

The volume of water on the face of the Earth is so huge, that the 'quota' per living person amounts to something like the capacity of Lake Windermere (330 million tons).

Man, however, is so modest in his consumption of water – averaging only about 2.5 litres (0.6 gallons) per day – that even if the intake of a representative individual and that of all his forefathers back to the Australopithecus (one million years ago) were to be aggregated, only about 0.3% of the individual's personal 'quota' would have been used.

... all his forefathers back to the Australopithecus ...

Nasal Storms

The maximum speed of the air passing through the nose in the course of normal inhalation approximates wind force 2 of the BEAUFORT Scale (3.4 metres = 10 feet per second), described as a *light breeze*.

Divorcing

Four out of every hundred children in the United States have divorced parents.

Something about New York

In 1626 a Dutchman, PETER MINUIT, bought Manhattan Island from its Indian dwellers for 24-dollars worth of ribbons and beads. To raise the estimated real-estate value of its 77 km² today, each U.S. citizen (women and children included) would have to contribute six times the original sale price (a total of $25 000 million).

New York's spectacular showpiece, the Empire State Building, with its 449 metres (1472 feet) the tallest building in the world, is visited daily by some 50 000 people, one third of whom are employed in its offices. The seventy-eight passenger elevators have already travelled the equivalent of 500 times the circumference of the Earth. At the end of a working day, the Rockefeller Centre disgorges a population like that of New Haven, Connecticut or Heidelberg on to the sidewalks.

Excelsior

A Berlin architect has designed a single building to accommodate 25 000 people. This self-countained *city* would rise to three times the height of the Empire State Building, its highest point a full kilometre (3250 feet) above ground level. Not the least of the advantages of 'town planning' on this pattern would be the great reduction, if not the complete absence, of traffic accidents.

Sense Organs

The World of Sound

The science of acoustics is a field full of surprises to those unacquainted with the physics of sound, and even such familiar transmitters and receivers of sound waves as our vocal chords and our ears conceal scientific truths of which we are normally but dimly, if at all, aware.

– Most sound-producing mechanisms are extremely inefficient, their output in terms of energy being only a minute fraction of the mechanical or other energy necessary to activate them*.
  Consequently, most of the sounds of our daily lives – the crying of a baby, the buzzing of a bee, the chirrup of a sparrow, the hum of a passing car – each represent powers almost inconceivably small in comparison with that issuing from such sources of radiant energy as a burning candle, electric light bulb or domestic stove.

– In view of these facts, it is astonishing that the human ear is capable of hearing such low levels of acoustic energy. Under favourable conditions, we may actually perceive sound waves with a power of only 10^{-16} of a Watt impinging on our eardrum so that this extraordinarily low value is a fairly accurate measure of our threshold of hearing. In fact, this threshold value is so close to the energy of thermal motion, that if our ears were only very slightly more sensitive, we might experience the constant background noise of the collisions of molecules in the air – a very tiring sensation!

* For the sake of simplicity we neglect the question of sound pressure.

– On the other hand, the human ear can endure, without damage, short-lived surges of sound 100 million million (10^{14}) times higher than the threshold value. The total output in acoustic Watts of such sound sources (only a fraction is caught by our ear) is comparable to the electrical energy output of a power station serving a small city!

Our threshold of hearing

Our eardrum and the cross-section of the outer ear have a surface area of approximately 1 cm², so that the power impinging on the eardrum in the presence of diffuse sound is only a very small fraction of the total power emitted by the sound source. Nevertheless, under ideal conditions of sound propagation (*i.e.* if no conversion of acoustic to thermal energy took place, and in the absence of wind deflection), the human ear could detect a sound source of 100 Watts (1/8 horsepower) at the distance separating Brussels from Athens, or Denver, Colorado, from Boston, Massachusetts, *i.e.* about 3000 kilometres (1875 miles)!

A walk through the world of Sound

Though our ideas about acceptable noise levels are generally very subjective, it is true to say that a ten-fold increase in the power of a sound source doubles the degree of loudness. If we found ourselves in an environment where the level of sound was at the threshold of hearing (10^{-16} Watts impinging on our eardrums), and we wished to visit five different localities in each of which we should be subjected to sound levels ten times greater than the last, our itinerary might be:

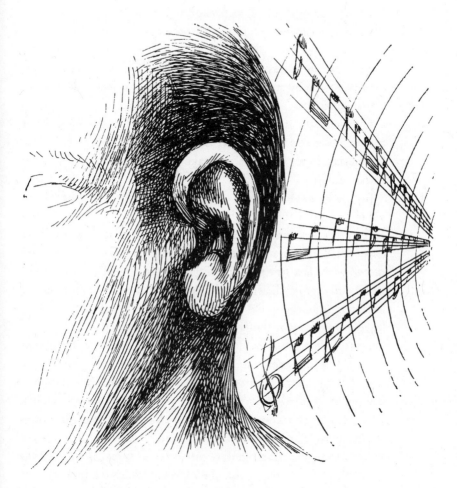

Just as we are inclined to think that ten million times the threshold of hearing is
the limit of our endurance ...

– a meadow where the grass faintly rustled in a gentle breeze;
– the centre of a deserted city park;
– a street closed to traffic;
– a large city square with its background hum;
– a vantage point some fifty yards from a busy motorway.

If our experiment has been well arranged, the noise level has now increased, step by step, to 100000 times the threshold value of our starting point.

If we now enter a large department store, our ears are perhaps again exposed to a ten-fold increase in sound level, and this rises, say, by another factor of ten as we make our exit on to a street in the throes of the peak-hour traffic.

Just as we are inclined to think that ten million times the threshold of hearing is about the limit of our endurance, the roar from a railway viaduct convinces us that we can stand at least a further exponential increase. In fact, the exhaust crackle of a powerful motorcycle a few feet away, and the staccato pounding of a pneumatic drill across the street, each of which represents a successive heightening of the sound level by the same factor of ten, prove to leave our ears momentarily deafened but still intact.

By this time, we have subjected our ears to levels of sound considerably beyond what might be called the *damage-risk limit*, i.e. the noise limit at which we may suffer permanent loss or degradation of hearing if exposed to it for long or regularly recurring periods. Nevertheless, the acoustic power concentrated on our eardrums by the pneumatic drill (10000 million times the threshold value) may still be exceeded by ten thousand times before we experience the so-called *threshold of pain*, i.e. the noise level at which we immediately experience severe pain in our ears, and to which we should, for reasons of safety, never allow ourselves to be exposed for more than a few moments. Airline ground staff, who would otherwise be subjected to this noise level in the close proximity of modern jet aircraft, are instructed, and are certainly wise, to wear earplugs. Even this extreme level of sound, however, is negligibly small in terms of energy output when compared to the power consumed by ... a pocket flashlamp.

Full of Sound and Fury, signifying... very little

The sound energy developed by a large concert orchestra in the course of a two-hour programme is about equivalent to the energy used by the conductor in covering the last six steps to his dressing room. In terms of mechanical energy, even a hearty applause from the entire audience is a very poor token of appreciation of the orchestra's performance; it corresponds to something like the energy expended by one member of the audience in raising an arm up to his head.

The power output of the sound produced by a man who talks for three hours a day throughout an average life-span is roughly sufficient to heat a cup of tea up to the temperature at which it is normally drunk! In other words, if the speaker's food consumption is 3000 kcal/day, his lifelong oratory will absorb the energy equivalent of only a couple of slices of bread and butter!

If all the inhabitants of the world were to speak at the same moment, the total acoustic power produced would be comparable to the output of a rural power plant (0.3 Megawatt).

In order to produce a level of sound approximating that of a modern jet engine operating at full thrust, the whole world would need to shout about three times louder than a normal speaking tone whilst the total volume was concentrated at a single point.

The human being is thus a poor 'loudspeaker', but has a remarkable capacity for enduring extreme levels of sound. In view of this capacity, it is apparent that the displacement of the eardrum from rest in response to threshold values of sound must be almost immeasurably small. In fact, it is of the order of one tenth the diameter of a hydrogen molecule, *i.e.* one thousand millionth (10^{-9}) of a centimetre!

Combating Noise

Now that jet aircraft with their accompanying inferno of sound are a part of our everyday experience, and with the prospect of their number increasing, it is comforting to learn that the problem of noise-suppression has its redeeming features.

Jet-engine combustion produces two sorts of noise: the primary noise due to the whirling flames of the burning gases, and the secondary noise caused by the tenement or crest of the flame acting as a sounding-board for the flow and vibration of the gases.

The promising fact is that the primary noise, which is the more difficult to suppress, is much the weaker of the two. In fact, though the scream of a modern jet engine exceeds the threshold of pain (as described above), the primary noise is hardly more intense than that produced by an underground train – quite acceptable to the commuter!

Sight and Sound

Faced, say, with the surgical necessity of accepting either blindness or deafness, most of us would presumably resign ourselves with greater reluctance to darkness than to silence.

Social and psychological considerations apart, however, our faculty of sight is functionally far inferior to our powers of hearing, for whereas only about one octave of the frequency spectrum of electromagnetic waves (between the wavelengths 3800 Å to 7600 Å) is visible to the human eye, at least nine octaves of the spectrum (between wavelengths of 3 cm = 1.2 inches and 10 metres = 33 feet) are audible to our ears.

The Threshold of Vision

The point at which the average human being is first able to perceive a faint source of light is called the *threshold of vision*. The amount of radiation necessary to palpably stimulate the human optic nerve is so infinitesimally small, that if the mechanical energy required to lift a single pea through one inch were converted to light energy, this would provide stimulus enough to have actuated the optic nerves of every human being who has ever lived.

Delicate Taste

The threshold of acceptance below which animals refuse food varies very greatly. For example, most bees will accept a 20% sugar solution, but refuse* a concentration of 10%.

There is also a threshold of perception, which is determined, in the case of bees, by first starving the creatures, and then giving them the choice between solutions of sugar in water of different concentrations.

---

* In the case of the bee the reason for its refusal may be that solutions of low concentration would not yield a honey that would keep in the hive through winter: the sugar is the food, but at the same time the anti-freeze. Diluted solutions would also mean uneconomic transportation.

Although the bee nourishes itself almost exclusively on sugars, it is ill equipped to find them, its threshold of perception being rather elevated (3% concentration for sucrose). It is quite astonishing that in man, for whom sugar is only one of several important foodstuffs, the threshold of perception of sucrose is seven times lower.

The minnow, which tastes with its whole skin, has a threshold of perception 400 times more sensitive than that of the bee, whilst the Red Admiral butterfly, with taste organs in its feet, excels the bee's capacity by no less than 800 times.

The Rhythm of Consciousness

Suppose we were able, within the length of a second to note 10000 events distinctly, instead of barely 10, as now. If our life were then destined to hold the same number of impressions, it might be 1000 times as short.

We should live less than a month, and personally know nothing of the change of the seasons. If born in winter, we should believe in summer as we now believe in the heats of the Carboniferous era. The motions of beings would be so slow to our senses as to be inferred, not seen. The sun would stand still in the sky, the moon be almost free from change.

But now we reverse the hypothesis and suppose a being to get only one thousandth part of the sensations that we get in a given time, and consequently to live 1000 times as long.

Winters and summers will be to him like quarters of an hour. Mushrooms and the swifter-growing plants will shoot into being so rapidly as to appear instantaneous creations; annual shrubs will rise and fall from the earth like restlessly boiling water springs. The motions of animals will be as invisible to us as are the movements of bullets; the sun will soar through the sky like a meteor, leaving a fiery trail behind him (M. ČAPEK).

Energy

Energy Resources—Present and Prospective

The Sun, our main source of Energy

In transmuting hydrogen into helium, the Sun continuously converts so much matter into so vast an amount of energy that we can hardly comprehend the magnitude of the power liberated: 3.5×10^{26} kcal/hour.

The weight of matter converted into energy and lost to the Sun per second is equivalent to the freight capacity of about 400 ships each of 10000 tons.

Just for the purpose of a first orientation in this field we mention that in terms of human consumption this energy output means the following. If mankind were able to completely harness the Sun's energy, each of us could have at his or her disposal 70000 times the total power capacity of the U.S.A. (160000 times that of Western Europe).

The weight lost to the Sun per second is equivalent to the freight capacity of 400 ships each of 10000 tons.

The Earth, a humble receiver and a poor storehouse

Due to its great distance from the Sun, the Earth intercepts only a tiny fraction (1.5×10^{17} kcal/hour) of its supplier's total output. Even so, if we could learn to utilize this still prodigal flood of energy in a perfect way, wide prospects would be opened up.

Man's present average daily calorie consumption per head being the equivalent of about 5.2 kg (11.4 lbs) of coal, can be represented as follows:

– 0.36 kg of coal (2500 kcal) as food,
– 4.8 kg of coal (33 000 kcal) as power.

On this basis of food and power consumption, solar energy could satisfy the needs of so many people (1.2×10^{14}) that each two square metres (22 square feet) of all the land available on earth, could support one man.

Since few of us would wish to find our neighbours so near, we will put this figure in a more attractive way: each of us could avail himself of as much food and power as is now consumed by 6500 U.S. citizens (16000 Western Europeans).

In reality, however, we make but little use of the stream of solar energy which bathes our planet. Only one percent is directly involved in supporting life (man, animals, plants) – the remainder is either reflected or it is absorbed and later partly emitted.

Many of us are perhaps inclined to forget the great reserve of energy in the form of sub-terranean heat, very high temperatures prevailing 1000 miles or so under our feet. The outward flow of heat escaping at the Earth's crust alone is estimated to exceed our total calorie requirements by a factor of ten, but due to low temperatures it is admittedly extremely difficult to exploit. Unfortunately(?), only about one percent of the outflow of heat is concentrated in volcanoes where the possibility of utilization is, perhaps, greatest.

The above situation is the more disappointing in view of the Earth's limited reserves of coal and oil. These are estimated to be equivalent to only two weeks' supply of solar energy poured on the Earth's surface, and they will be exhausted (or at least no longer available at present cost) within a couple of centuries or so.

Consequently it is obvious that energy sources, other than coal and oil, must be exploited if mankind is to survive. We will come back to this point after having considered the potentialities of solar energy in the future.

Harnessing solar energy for our heat and power requirements

Up to the present, man has been successful, to some extent, in satisfying his heat and power requirements from the daily stream of solar energy: indirectly, by the combustion of wood and the remains of plants and animals, directly from the dynamic forces of water or wind and from the Sun's radiation. It is estimated, however, that this way of utilizing solar energy will come to provide no more than 15% of man's needs in the course of the next century, since the world population is increasing much more rapidly than the economic exploitation of this kind of the solar energy can be expanded.

It is theoretically true that mirrors arranged to concentrate solar energy could supply two-thirds of the power requirements of our 21st century descendants, but such an arrangement, if at all feasible, would certainly have to be confined to the tropics, *e.g.* for cooking purposes. In the more temperate regions it would be necessary to cover about half the surface area of industrial countries with such mirrors to attract the required energy, whilst vast accumulators would be needed to store energy against cool and cloudy days.

Despite the apparently simple technology involved, wind power could economically meet only about 0.1% of future power requirements. Even in The Netherlands, where wind power has a long tradition of useful service, windmills would have to be erected along the entire coastline, with their sails capable of catching all the wind up to a height of some 160 metres (500 feet), before the country's present power needs could be satisfied from this source.

Harnessing solar energy for food production

We saw already that man's present intake of food per head has an average value of 2500 kcal/day, which corresponds to about 1/3 kg of coal or (to express it in a more appetizing fashion) twice this quantity of sugar. Expressed in terms of solar energy, we can say that in order to feed himself, man makes use of only two out of every million calories poured by the Sun over the Earth.

The reasons for this extremely inefficient use of the vast amounts of solar energy available for man's nourishment are threefold:

– relatively very little of the Sun's energy reaching the Earth is taken up by plants and animals (much is absorbed by the atmosphere);
– man consumes an extremely small proportion of the Earth's vegetation and animal population;
– both plants and animals perform very poorly in fixing solar energy. For example, the quantities of carbon photosynthetically fixed from the air by plants corresponds to a mere 0.2% efficiency in assimilating solar energy.

On the latter point, however, there is potential enough for philosophical speculation. Controlled biological photosynthesis, exploiting the mechanism of certain algae (chlorella), could increase efficiency to 10%. This would mean that the acreage now given over to crops, if inundated to shallow depths, could adequately supply one thousand times the present world population with food.

Seeing that the harvests could be partly eaten and partly burned as fuel, the same acreage could supply both the food and power requirements of the world beyond the year 2000, by which time the world's population is expected to be nearly double its present size. Remarkable as it may seem, the volume of water required for this boring one-crop agriculture would actually be less than the amount now needed for the world's productive farmland with its excessive surface evaporation (leaves).

The Energy reserves on Earth

Energy from Coal A handful of coal (1/5 kg or 0.45 lbs), or a glass of oil (0.16 litres = 0.04 gallons) represent enough energy to lift a fully grown man from sea level to the height of Mount Everest (about 8900 metres or 29000 feet). The entire population of the world could be raised through the same distance on only two hours' output from all the world's coal mines.

However impressive the amount of energy releasable from a small amount of these fuels may be, human consumption is so large that – as mentioned above – the reserves will soon be exhausted and for the sake of our great-grandchildren alone we will have to find other energy sources.

Energy from Matter The energy available from the combustion of coal and oil fades into insignificance when we contemplate the energy potential of Matter (EINSTEIN's equation $E = mc^2$). Assuming that we were able to fully exploit this unlimited source of power, the following tasks could be performed on the energy released from one single gramme of matter:

– mechanically: lift the population of Turkey or Argentina (15 million) on to Mount Everest;
– thermally: boil a gallon (about 4 litres) of water for all the citizens of the United Kingdom (55 million);
– electrically: supply current to a city of 15000 people for a whole year.

In this context it is curious to reflect that:

– a passenger train could travel several times around the world on the energy represented by... a cardboard railway ticket!
– a large aircraft could operate continuously for a couple of months on power derived from... a breath of air!
– a large apartment could be heated or supplied with electricity for twelve months from the energy equivalent of... a handful of snow.

Energy from Uranium and fusion The fission of uranium 235 and the fusion of say hydrogen to helium are sources of energy intermediate between the combustion of coal and the complete conversion of matter into energy. Fission is being applied nowadays in atomic reactors. Hope exists that the process of fusion will some day be effectively controlled.

Though both processes are accompanied by the release of e-normous quantities of energy, they are still far removed from the ideal of complete conversion of matter to power, as an example may illustrate.

An electric power station with a capacity of 250 Megawatts, serving a city of a good half million people (Newark, USA or Bristol, UK) could be run* at full capacity by:

– 10 kg (22 lbs) of coal for   1/3 second
– 10 kg of uranium-235 (fission) for   13 days
– 10 kg of hydrogen (fusion) for   3 months
– 10 kg of matter (conversion) for   50 years

Unfortunately the amounts of uranium now available are not sufficient to offer a final solution to mankind. If the age of the Earth

* Taking the efficiency of the power plant at 33%.

were one-tenth greater than it is now estimated to be, uranium 235 (on which we increasingly rely as a reactor fuel) would not be available to us; radioactive decay would by now have led to its extinction. As it is, present residual amounts of uranium (and thorium) will make it possible for us to have sufficient energy available for only a few centuries at reasonable cost*.

Outlook Consequently human society cannot afford delay in developing new, economically exploitable sources of energy, whether it be fusion, the conversion of matter into power, or... the humble chlorella: the first two sources could supply energy in practically unlimited quantities.

But even the layman should be aware that it is quite unrealistic to look at present energy reserves (apparently sufficient for 4000 years!) whilst forgetting that the prodigious growth of population and *per capita* consumption will reduce their effective life to at most a few centuries.

Fuel and Power

Already aware of the fact, stated in the foregoing article, that a handful of coal can theoretically provide enough energy to lift a man to the top of a high mountain, our engineering friend Ernest has meanwhile been calculating what level of efficiency in lifting things we have actually achieved in our technical age. His conclusions are hardly encouraging.

Taking the summit of Mont Blanc (4800 metres = 16000 feet) as the experimental goal, he finds that 100 g or so of coal ought to be quite sufficient to transport our man to the top, *i.e.* at 100% efficien-

---

* By using quick neutrons, this period might be considerably extended.

cy. It then appears that railway locomotives were invented for the primary purpose of burning fuel as wastefully as possible, for in transporting, say, 1000 passengers to the summit, their actual or equivalent consumption of coal would be no less than fifty to eighty times the theoretical optimum:

– diesel or diesel-electric locomotive: 5000 g per passenger,
– steam locomotive: 8000 g per passenger.

Ernest now turns his attention to the petrol engine and concludes that in the motor car we have certainly found a more economical means of transport. The amount of petrol necessary to afford four persons a comfortable view from the top of Mont Blanc turns out to be equivalent to 3500 g of coal per head – a considerable improvement, of course, but still 35 times the theoretical optimum.

Our intrepid engineer then discards his overalls and sets himself to climb a vertical ladder to the summit. Consuming only the equivalent of 850 g of coal from the lunch-box in his rucksack, he arrives at the top panting no more vigorously than the steam locomotive. His efficiency has thus been $12\frac{1}{2}\%$, and his climb has required hardly one-tenth of the fuel devoured by the locomotive from its tender in bringing a single passenger to the same point.

Ernest, however, is vexed by the thought that he has had to drag the whole weight of his body (70 kg = 150 lbs) up the mountain. He therefore constructs a hoist with a counterweight which balances the cage and a fair proportion of his body weight, and he is now able to reach the summit while expending the equivalent of only 400 g of coal – 12 times more efficiently than the train-borne traveller… and a good deal more quickly!

Ernest has certainly started something! He and his professional colleagues will be kept busy installing lifts in practically every new tall building. Those who continue to use the stairs presumably need the exercise; they burn sixteen times as many calories doing it the 'hard way'!

Equivalence of Energy and Mass

If a passenger car with a stationary weight of 1000 kg (2200 lbs) is accelerated from rest to 60 km per hour (40 miles per hour) it gains something like the weight of a pinhead in the process. If the car could be made to travel at one hundred times the velocity of sound (100 × 1200 km per hour) it would become about 100 kg heavier, and at 250000 km per hour its weight would be doubled. Travelling at 0.999% of the velocity of light, the car would weigh two thousand times its stationary weight, and would plough deep furrows in the surface of the road.

Hard and Light Work

The bricklayer who climbs a steep ladder carrying a load of 23 kg (50 lbs) on his back, performs a particularly arduous task – his consumption of energy is about 20 times what he expends when lying on his back, *i.e.* 20 × 1700 kcal/day.

Walking on loose snow with a similar burden, or rowing at the rate of 33 strokes per minute, also demand very considerable effort, requiring about 75% of the work involved in climbing the ladder.

Even the apparently trivial activity of dressing increases one's energy consumption to about 3½ times that of the supine state; walking at a speed of 4 miles per hour (6.4 km per hour) is only twice as exacting as dressing.

In terms of physical effort, watch repairers and draughtsmen should have an easy time of it – in the normal pursuit of their vocations they probably exert themselves only half as much as they do in the course of dressing.

Not many people appreciate the often heavy physical exercise the house-wife, too, has to perform. Making the bed is four times as tiring as lying in it.

However, there are also psychological factors involved. Just imagine her complaints if, instead of enjoying a dance for an hour in the evening, she had to make beds for three quarters of an hour (with the same periods of rest in between) or was required to be her husband's sparring partner for half an hour! Dancing requires three times the energy for the supine state, making beds four times, boxing seven times.

Some Technical Facts

The Welding Family

The *old-fashioned* welder, armed with an oxy–acetylene torch, can produce a heat concentration of 10000 Watts per cm².

His son, the electric welder, can achieve ten times this value.

The grandson, making use of the electron beam, is able to surpass his grandfather's performance by about one hundred times.

Finally, in the hands of the great-grandson, the laser (see page 71) is capable of developing temperatures far above the boiling point of any element, producing a heat concentration of the order of 1000 million Watts per cm².

... his son, the electric welder, can achieve ...

Optical Illusion?

Among the feats ascribed to the great Greek mathematician and natural philosopher ARCHIMEDES is his apparently successful attempt to fire a ship of a hostile Roman fleet by means of a mirror. It is difficult, however, to reconcile the episode with the pertinent laws of physics, on the one hand, and with the military technology of his opponents, on the other.

In the, in any case, unlikely event that ARCHIMEDES happened to have a 10-foot (3.3-metre) diameter concave mirror at his disposal, he could conceivably have set a wooden ship on fire over a distance of 100 feet (33 metres), since such a mirror could concentrate the Sun's radiation to a hundred times its normal intensity. At such a limited separation from the target, however, the mirror and those manipulating it could hardly have survived unscathed the hail of arrows, spears, stones and the like with which the Romans would surely have defended themselves. On the other hand, if the mirror remained beyond the range of the farthest-carrying Roman missiles, its effectiveness as an incendiary weapon would have been very limited. At 1000 feet, for instance, the reflected radiation from the mirror would have been hardly more intense than direct sunlight.

Deterrent

The total explosive power of nuclear devices detonated since 1945 corresponds to something like ten thousand times that of the atomic bomb dropped on Hiroshima. Fortunately, these devices were used experimentally, otherwise they could have annihilated three or four times the population of the United States.

Smooth Talk

The metallurgist is not easily impressed by the poet's *burnished brass* and *flashing steel*. He will tell you that, under the microscope, the surfaces of the polished axles of your car, turning in their polished bearings suggest nothing so strongly to the imagination as the encounter of Alps with Andes.

... the surfaces of the polished axles of your car ...

The Atomic Tennisball

To equal the explosive force of an atomic bomb containing a charge (fissile material) the size of a tennisball, a quantity of dynamite would be needed sufficient to pack Times Square (New York) or Piccadilly Circus (London) to the height of the surrounding buildings.

Pure Materials

Uranium for use in atomic reactors should ideally be refined to such a degree of purity that not more than one borium or cadmium atom remains for every ten million uranium atoms, corresponding, say, to one albino in the entire population of New York City.

Certain types of transistor require an even stricter standard of purity, the admissible inhomogeneities in the finished product being comparable to a situation in which only tree albinos were represented in the world's total population, *i.e.*, 3 : 3 000 million.

A Cracking Pace

A crack produced through stressing rigid materials like steel or glass progresses with a speed of about 2 km per second, roughly seven times the speed of sound (in air). If the Earth's crust possessed the characteristics of such materials, a crack would travel round the entire circumference of the world within six hours.

Wasting Away

Wear and friction can be the most insidious enemies, their assaults being often effective far beyond their apparent significance. For example, a large truck consigned to the scrap-heap may have lost no more than 0.1% of its original weight, and yet the pound or two removed by chemical or mechanical attack have rendered it unserviceable.

A similar, apparently insignificant degree of wear gradually frays the collar and cuffs of your shirt, and causes you to discard a garment which is otherwise, perhaps, *as good as new*.

... consigned to the scrap-heap may have lost no more than 0.1% ...

Buildings on the Move

Observatories are equipped with extremely delicate measuring instruments... with which some remarkable observations have incidentally been made with respect to the observatories themselves.

At one establishment it was discovered that the building inclined 1/10 mm towards the East in the morning and leaned towards the West to a similar degree in the evening, from which it was apparent that large buildings in general *follow* the course of the Sun to a measurable, if extremely limited, extent.

Such observations have disclosed the curious fact that the observatories of Leipzig and Rome have moved a finger-length closer together over a period of sixty years, whilst in the course of eighty years the observatory at Heidelberg has become a hand's length more remote from the North Pole.

Vacuum

If 1 cm^3 of air were to be spread evenly over a surface equal to the area of the U.S.A., there would be about 250 molecules to every space the size of a fingernail (say 1 cm^2).

High-pressure techniques can achieve pressures of the order of 100000 atm. Even distribution over the area of the U.S.A. of 1 cm^3 of air under such compression would give each *fingernail* a population of 25 million molecules.

Modern ultra-high-vacuum techniques allow the number of molecules in a confined space to be reduced to only 4 per cm^3, which closely approximates the extremely low densities found to obtain in interstellar space.

At present the dimensions of ultra-high vacuum chambers range up to only a foot or two (60 cm), but larger ones are being designed to test the behaviour of even complete aircraft.

As can be expected, it becomes extremely difficult to eliminate minute leaks of air into chambers of this kind. The designers of this advanced hardware regard as 'intolerable' a leak permitting the entry of only one cubic centimetre of air per century. However small this quantity seems to be, the number of entering molecules then is still something like three times the world population per second!

Inland Sea

Whenever the sluice-gates of the large sea-lock at Ymuiden, The Netherlands, are opened to allow a ship to pass, a quantity of salt sufficient to fill between fifty and one hundred railroad freightcars is introduced into the interconnecting waterways.

Light Conversation

The laser is a device which effects *l*ight *a*mplification by *s*timulated *e*mission of *r*adiation. By *pumping* (stimulating) the electrons in certain materials (crystals, gases, semiconductors, etc.) from lower to higher energy levels, and allowing the resulting preponderance of electrons at the latter levels to discharge themselves *coherently* (in unison), a very powerful, extremely narrow beam of light can be produced. Such a beam, if directed at the moon (385 000 km = 240 000 miles away), is so 'narrow' that it would throw a circle of light with a diameter of 'only' some 80 km (50 miles) on the lunar surface.

Provided suitable means of modulation can be found, it has been estimated that a single laser beam could be used to relay of the order of 100 million-simultaneous telephone calls or... all the radio and television programmes being transmitted at any given time.

More Regular than Clockwork!

By applying the maser (*M*icrowave *A*mplification by *S*timulated *E*mission of *R*adiation) principle, cesium crystals can be made to produce resonance vibrations with an extremely high degree of constancy. Regulated by the tempo of these vibrations, the atomic clock will lose no more than one second over a period of 300 years.

Fast Photographic Films

Modern high-speed film can be as much as 1000 million times faster than ordinary kinds of photographic film. On an extended time scale, a one-second exposure time for the former would correspond to about 30 years for the latter.

High-Speed Cinematography

High-speed cine cameras, based on the rotating-mirror principle, commonly employ mirror speeds of 3200 revolutions per second, with a frequency of 50 million frames per second. Insurmountable practical difficulties apart, such cameras could make separate photographs of all our contemporaries, standing shoulder to shoulder, in roughly one minute.

Science and Education

Brains and Machines*

With the modern computing machine threatening redundancy to a growing number of human calculators, and with today's 'electronic wonders' themselves facing obsolescence as each new 'generation' of computers leaves the drawing board, it is interesting to compare the performance and capacity of the human brain with the present capabilities of its most sophisticated brainchild.

– The first point of comparison should, perhaps, be the *degree of complexity* – a fundamental property, because a ten-fold increase in it can convert a machine capable, say, only of *addition* and *multiplication* to one able to cope with *division, raising to powers* and *interpolation*. In this respect, the brain is without doubt vastly superior, possessing more than ten thousand million cells, whereas a computer with 'only' a million transistors already presents considerable problems to the electronic engineer. It is safe to say, then, that the brain of a single human being will long remain the equal in complexity of several thousand computers taken together.

– Under the heading of *degree of complexity* can also be included the concept of *memory*. In order to store the basic multiplication table in the vacuum tubes of a computer, 1500 binary digits or *bits* are required. Concerning human memory, estimates vary very widely as to its capacity for retaining information, some putting it as high as 60000 items, each as complex as the multiplication table. By contrast, a modern computer's storage capacity of 3 million bits corresponds to a *memory* able to recall only 2000 such items, which means that the human memory may well surpass that of even an elaborate computer by a factor of a thousand or so.

* Partly condensed from J. G. Kemeny.

It is astonishing to realize that if the human memory is indeed capable of retaining 60 000 complex items of information (10 million bits), it must acquire the bits at an average of one every twenty seconds!

– The human brain can receive impressions from the world outside it and make independent correlations between them. A man deprived of his five senses would be comparable to a computing machine with a fixed tape, for such machines can only *imitate*. Computers must be programmed by man; man's brain programmes itself.

– In the matter of *economy of energy*, too, the human brain far excels the computer. The entire brain, with its myriads of cells, operates on less than 100 Watts, whilst a modern computer would consume the power output of a generating station designed to serve a community of 200 000 people, if it had to activate a like number of cells (transistors).

– On all the above counts, then, the computer is certainly no match for the human brain; but what are the compensatory advantages which have won it so important a place in the world of commerce, industry and science and given it so vital a role in military defence planning?

Pre-eminent of its virtues is *speed*. The computer can already operate up to ten thousand times faster than the brain, for whilst the same nerves cannot be 'triggered' more than about one hundred times a second, the vacuum tubes of the computer can be switched on and off perhaps a million times in the same period. And before very long separate computer operations will come to be measured not merely in millionths, but in thousandths of millionths of a second: nanoseconds... of which, as remarked elsewhere in this book, as many pass per second as seconds pass in thirty years!

... that the brain of a single human being will long remain the equal in complexity of several thousand computers taken together ...

When one considers also that computers need not suffer from fatigue, and will not normally make mistakes, their advantages are quite apparent. A few examples of their time-saving capacity will demonstrate how seemingly limitless their possibilities are.

– Some modern computers can execute 3 million operations per second, which means that in under twenty minutes they could complete a calculation involving some item of information about every individual human being on Earth!

– A fast computer could perform in roughly ten minutes all the calculations which the average clerk would normally reckon to carry out in the course of his entire career!

– During the last century a Danish schoolmaster calculated the value of π to eight hundred places of decimals – the feat occupied his whole life. Recently, a medium-size computer took a few hours to check his figures... they were free of error!

– A multiplication sum like $875,426,319,764 \times 183,914,632,284$ can be completed by a computer-robot in less than five seconds. Sums with only nine figures on either side are child's play to him: he can reel them off to the tune of five thousand per second!

Registering Progress

The intensity of industrialization in a given country can be gauged to some extent by the number of patents taken out by its citizens. In the United States, for example, this number is now one for every sixty citizens, whereas the average for the world is one for every three hundred persons.

Faithful Figures

Of all the religions of the world, Christianity claims by far the greatest number of nominal adherents, averaging 26 persons per hundred of the total population, of whom 15 are Roman Catholic, 7 are Protestant, and 4 are Eastern Orthodox.

Mohammedanism, the second creed in numerical importance, is confessed by half as many souls (13) whilst Hinduism and Confucianism number 10 followers each per hundred persons, and Buddhism 7.

For all its outstanding contributions to Western art and culture and its eventful history, Judaism is represented by scarcely more than 12 million persons, corresponding, say, to a child in our sample of one hundred people (0.4%).

Of all the religions of the world ...

Lunacy in Research

In 1961 the cost of federal and private research in the United States (14000 million dollars) surpassed the value of U.S. automobile production by 2000 million dollars.

In the same year, space research absorbed a modest 1000 million dollars of taxpayers' money, but when an estimated 6000 million came to be earmarked for this research in 1966, the President was declared by some to be suffering from 'moon sickness'.

The first U.S. attempt to land a manned spacecraft on the moon, planned for 1970, is estimated to require 20000 million dollars, including all research and development costs. This means that, on the average, each U.S. citizen will contribute $150 towards the expense of this undertaking, amounting, for a family with two children, to the cost of a new car.

The Happy Innocents?

Seven persons in every ten of the world's population are not reached by the common means of communication (newspapers, radio, television, telephone) and are thus largely ignorant of the doings of their fellow men.

Lingua Franca

Though French is the mother tongue of only 2½% of the world's population, its continuing use as an international language seems assured – no less than 35% of the delegates to the United Nations express themselves *en français*.

Illiteracy

The fact that only half of the present adult population of the world can read or write is rightly a matter of grave concern to UNESCO (United Nations Educational, Scientific and Cultural Organization). This body has, moreover, estimated that no less than 2000 million dollars, spread over ten years, would be needed to provide the necessary instruction to two-thirds of the 500 million illiterates under its care.

Polyglot Extraordinary

If Cardinal MEZZOFANTI, keeper of the Vatican Library until his death in 1849, had been invited to deliver a one-minute address of welcome to an international conference in all the tongues at his command, he would have required over three hours to do so. This celebrated linguist is said to have mastered no less than 114 languages and 72 dialects (of the 4000 languages, spoken today).

Public Image

Despite the inroads television has made on the popular appeal of motion pictures, the world still numbers nearly a quarter of a million cinemas which each attract a daily average of two hundred patrons.

This is equivalent to the population of U.S.A. going daily to the movies.

Book Review

A well-known Dutch journalist, now in his sixty-sixth year, makes the following observations on reading habits in his country:

– of the four books bought on average by each of 12 million compatriots, three are belles lettres;
– one in every five books is bought for presentation as a gift;
– many people possess books which they have never read.

Of the 3000 volumes comprising his own library, the journalist finds he has read only 2000. He calculates that he would have to live to the age of 105 in order to do justice to the remainder... which does not dissuade him from buying yet more books!

Sic Transit...

For all the inspired creativity and exquisite crafsmanship of French and German artists of the past, their latter-day compatriots seem no less prone to forsake the old for the new than the less richly endowed nations of Europe.

In 1962, for example, the modern airport of Orly near Paris drew no less than 3½ million sightseers, whilst the Eiffel Tower attracted about half this number of visitors. By contrast, the incomparable treasures of the Louvre Museum and the majestic splendour of the Cathedral of Notre Dame could command the attention of only 25% and 5% respectively of the number of airport enthusiasts.

Crossing the Rhine in 1963, the statistician discovers that in the course of the preceding seven years the average German citizen has been but once to a museum, only twice to a theatre or concert hall, but no less than seventy times to a cinema. Small wonder then, that

German cinema proprietors gave public notice in their foyers of their indignation at the fact that, whereas the nation's theatres received state subsidies totalling DM 250 million ($60 million), the cinemas were obliged to pay DM 55 million ($15 million) in tax.

... once to a museum, only twice to a theatre, but seventy times to a cinema ...

Economy

Investing One Dollar

If, in the year 1800, NAPOLEON had invested $1 at 4% compound interest, the accrued capital could now finance a first-class round trip by plane from Western Europe to New York ($700).

If COLUMBUS had made a similar investment about the time of his first arrival in the New World (1492), a fortunate legatee would find himself possessed of 77 million dollars with which to make a transatlantic crossing in his own ocean liner (of about 30000 tons).

A magnanimous CHARLEMAGNE, investing a dollar on the same terms in the year 800, could have decreed that his present-day executor disburse to each member of the human race the sum of $10000 million (the declared capital of the Standard Oil Company of N.J. – Esso).

A dollar invested at 4% compound interest in the year of Christ's nativity would now have the value of roughly one hundred thousand globes of solid gold, each the size of the Earth! At simple interest however, the investor would now be entitled to withdraw about $80.

More Precious than Gold

The expression 'more precious than gold' may well lose its currency before very long. Weight for weight, uranium now costs fifteen times, and plutonium thirty times, as much as gold. However, even these radiant commodities pale before the magnificence of the neutron, worth a million times its weight in gold by virtue of its extremely costly production.

Steel, once a metal of some nobility, now cuts a rather poor figure. One gramme of neutrons could buy enough to build a ship of 15 000 tons.

Standards of Living

Recent statistics throw some light on the respective standards of living of the Soviet and U.S. citizen.

The Russians seem to be less talkative (by telephone), they occupy fewer rooms and are less mobile (by car) than the Americans. Although their meat diet is lighter, they seem less healthy.

Per thousand citizens the following comparisons can be drawn:

Per 1000 citizens	U.S.S.R.	U.S.A.
Telephones	20	400
Rooms occupied	650	1400
Automobiles	3	300
Meat consumed per year (equivalent in carcasses)	5	10
Number of doctors	18	13

The Private Economist

A septuagenarian, looking back over his past, could reflect that:

– if he had taken 10 minutes less to wash, shave, dress and breakfast each morning, and been quicker by the same margin in performing his daily preparations for bed, he could have utilized almost a whole year in less prosaic activities;

– if he had omitted from his daily fare one roll or slice of bread, his stomach would have had some 2600 lbs (1200 kg) less to digest;

– if, from his 20th year, he had regularly smoked a cigarette or two less than usual (representing, say, a daily economy of 6d or ¾ dime), he would have saved in the region of £500 or $1300;

– if he had parked his car one mile from his office, and walked the rest of the way in both directions each day, he would have indulged in healthy exercise to the tune of 20000 miles (30000 km), corresponding to 36 holes of golf a week.

Paid Aid

To increase the average level of earnings in the underdeveloped countries of Africa, Asia and Latin America by 2% – which would simply mean two ragged shirts rather than one for the vast majority of labourers – the annual total of wages and salaries would need to rise by some $20000 million. Unfortunately, the very low productivity of these countries places only one quarter of this sum within their capacity in the near future; the remaining three-quarters can only be provided from outside.

Over the last seventeen years the United States Treasury has spent over $35000 million in foreign aid: this considerable sum has contributed no more than 10% of the capital which would be necessary to implement a 2% rise in the general standard of living of the poorest nations of the world.

Swords and Ploughshares

The amount of money invested by the highly industrialized countries of the world for military purposes alone is estimated to be approximately equivalent to the value of the aggregate gross national product of the underdeveloped countries.

Productivity

In 1962 the 40 million farmers in the Soviet Union produced fewer cereals than the 6 million farmers in the United States.

Unequal Shares

Half of the population of Greece consists of farmers who together contribute only 0.5% to the country's tax revenues. Indeed, outside Athens there are scarcely four thousand families with incomes above the value of $4000. Such eloquent examples of unequal division of wealth can be found in many countries where industrial development is slow and unbalanced.

Uncommon Market

Until French recalcitrance put an effective stop to Great Britain's bid for membership of the European Common Market, New Zealand farmers were greatly concerned about the effect British partnership with *The Six* might have on their vital exports of agricultural produce to the U.K.

It would, no doubt, have been small comfort to them to learn that

American visitors to New Zealand spend on average the cash equivalent of two hundred crates of potatoes and one hundred fat lambs.

The Richness of the Dead Sea

The Dead Sea is estimated to contain about 20 tons of valuable salts for every man, woman and child on the Earth. The Great Salt Lake, though greater in size, is less concentrated, and could yield perhaps half this quantity.

Calorie Basket

An office clerk in the United States produces a daily average of 2 kg (4.4 lbs) of waste paper. At the end of each day, his garbage bin could thus provide more calories than he and his wife derive from their normal complement of meals.

Transplantation

Since World War II, synthetic-rubber plants have assumed a vital place in industry. If the present world production of synthetic rubber were to be discontinued, an area of agricultural land capable of feeding twice the population of Australia (26 million) would have to be given up to the cultivation of natural rubber.

Two Cars per Baby

At present in U.S.A. twice as many cars per year are produced as babies are born.

Prosit!

World production of wine is now sufficient to allow every member
of the human race to drink two gallons (nine litres) a year.

With Cola Compliments

The 1964 advertizing budget of one of the Cola companies is believ-
ed to have amounted to about 35 million dollars. Putting the intrins-
ic value of a bottle of Cola at a few dollar-cents, we may roughly
calculate that this company could have promoted its product at no
greater cost by giving a free sample to every family in the world.

Transport and Communication

Around the World

In the time of NAPOLEON and GEORGE WASHINGTON man's swiftest means of travel was the horse. Overlooking the natural barriers of sea, mountain and other limiting factors, a round-the-world relay race on horseback, using fresh animals to maintain maximum speed, would have taken thirty days.

At the speeds developed by the antelope, the carrier pigeon or an expert skier, the corresponding distance could be covered in about sixteen days, compared with the mole (walking, not burrowing) and certain kinds of snake, which would need two and four-and-a-half years respectively.

Assuming no land barriers to their progress, the flying fish might expect to complete the journey in five weeks; the salmon, in roughly half a year. Unhampered by headwinds, nor assisted by a following breeze, the swallow might arrive back at its point of departure in twenty-five days. This rate of travel was still far beyond human ingenuity in 1850, as JULES VERNE's Phileas Fogg discovered. His eighty-days' journey was indeed about the limit of possibility then, but was certainly far from being a practical solution to the problems of transportation of his day.

In modern times we are accustomed to rapid transit in a wide variety of forms, but scientists were recently surprised to find that radioactive clouds from a nuclear explosion circled the globe in only twenty-five days, i.e. with the speed of a swallow. Even more astonishing has been the discovery that tidal waves can travel around the Earth in a matter of sixty hours.

Not so very long ago it was popularly supposed that man could hardly hope to travel at much more than the speed of sound, which

... yet it is interesting to calculate that a steady walking pace ...

means thirty-three hours to complete the trip around the world.

The idea of humans bettering the fourteen hours a rifle bullet would theoretically require to cover such a distance seemed barely credible to many, until GAGARIN was rocketed into a circumterrestrial orbit requiring only 1.8 hours for each complete revolution. However, even this prodigious achievement pales before the ultimate, which can be set at the tenth of a second or so which would elapse before a beam of light, made to follow a curved path over the Earth, returned to its point of origin.

In the face of such extreme speeds the capabilities of our own two feet seem hardly worthy of regard. Yet it is interesting to calculate that a steady walking pace, maintained day and night, would be sufficient to cover the 40000 km (12000 miles) of the Earth's circumference between 1st January and Christmas Day of the same year.

... and Christmas Day of the same year.

Cheap Space Travel

Once the astronomical research and development costs of U.S. space programmes (see page 78) have been met by the American taxpayer, *national* expenditure on individual flights into space should prove relatively insignificant.

Circumterrestrial orbital flights of the kind undertaken by JOHN GLENN cost in the region of 3 million dollars, which represents a mere two cents out of the average American pay packet.

A probe involving circumnavigation of the moon is five times as costly, but still adds only the amount of a city bus fare (one dime) to the average taxpayer's burden.

An actual landing on the moon would, perhaps, double the Space Administration's outlay, but prestige-conscious American citizens will not baulk at the added price of a modest drink (2 dimes). They may even cheerfully pick up the tab for lunch (two dollars) when their Uncle Sam needs a cool 300 million dollars to take a closer look at Venus or Mars.

Driving on Alcohol

Though the use of alcohol by drivers on the road is rightly abhorred, and is certainly responsible for a significant proportion of highway accidents, there is reliable medical evidence to show that a healthy adult is capable of 'burning' about 10 cm³ (0.34 fluid ounces) of alcohol per hour. Consequently, a whole (standard) bottle of sherry or vermouth could theoretically be consumed by a driver in the course of a 15-hour journey without seriously impairing his faculties, provided he drank equal amounts (about one small glass) each hour.

The average alcohol content of cognac and whisky is, perhaps, double that of the wines mentioned above, so that the driver who permits himself to drink a whole bottle of such spirits, should have a journey of at least 30 hours before him – a very fatiguing prospect of itself, which immediately suggests the inadvisability of even 'touching a drop'.

France on and off the Eiffel

The total number of visitors who have ascended the 300-metre (980-feet) Eiffel Tower in Paris since it was opened to the public in 1889 now exceeds the population of France.

Police reports show that in the 75 years of the Tower's existence, 340 persons have jumped from it to their deaths. If the whole population of the world had climbed the Tower and the same proportion of visitors had taken this fatal decision, the horrifying plunge would have been a daily spectacle.

Absent-Minded

The traveller who today inadvertently steps on to the wrong train is usually able to rectify his error without too much expense or delay. For the astronaut of the future, however, the consequences of such an oversight could be far more serious.

Consider the case of the space traveller who intended to travel to the moon by express rocket (flying close to the velocity of light) and subsequently discovered that he was aboard the suburban transsonic capsule. Instead of arriving at his destination within one second, he would take nearly three weeks over the journey.

Incomparably worse would be his plight if the transsonic capsule proved to be bound for the sun. His little daughter would have finished her schooling by the time he had completed the outward journey, and would be a middle-aged woman of about forty years on his return!

... his little daughter would be a middle-aged woman ...

Rough Transport

King SENNACHERIB of Assyria employed one thousand slaves to draw a sledge bearing a massive carved stone. He could have achieved his object with the aid of only fifty of them if 'his engineers' had thought to provide the sledge with wheels – with which triumphal chariots were already equipped in his day.

The Killer Road

Traffic fatalities on the roads of Europe have now climbed to a twelve months' average of 65 000 – the population of Calais.

... on the roads of Europe ...

Inhospitable Outlook

If Rotterdam's present yearly toll of traffic casualties (excluding fatalities) were to be hospitalized at the same time, they would fill the city's sick wards to capacity for a period of six weeks. Appalling as this situation is, it seems unlikely to be an extreme example. Rotterdam has neither an exceptionally heavy traffic density compared with other cities of similar size, nor a dearth of hospitals.

Commuters

The modern tendency to completely separate the residential areas of large cities from offices, factories and other centres of commerce can give rise to considerable problems for the local labour force. In Germany, for example, it is a common situation to find only about 5% of the urban population actually resident in the business centre of a city, whilst upwards of 35% must make their way there every working day.

The present situation in Paris is such that the equivalent of half a million working days is 'lost' every day as a vast army of office employees, artisans and others travel to and from their places of work.

Busiest Port

The weight of cargo transshipped annually from one vessel to another in the port of Rotterdam is now equivalent to disembarking and re-embarking half the population of the world.

Airborne

In the fifty-five years since BLÉRIOT first flew across the English Channel (1909), air transportation has undergone phenomenal development. The number of passenger-miles now flown in twelve months corresponds to a flight over the Channel (31 km = 20 miles) once a year by every man, woman and child on the Earth.

... since Blériot first flew across the Channel ...

Busy Lines

The number of telephone subscribers in the world is currently estimated at 160 million, half of whom are in the United States.

The average number of telephone calls made by Americans works out to two for every citizen, five working days a week. The Canadians are even more talkative, it seems: their telephone conversations also average two per day, but they continue on Saturday.

Expensive Transportation at Home

When Baron LEO ROTHSCHILD installed a hydraulic elevator in his London house in 1870, it cost him more to travel between adjacent floors than to cross the entire metropolis by cab.

Many(!) Happy Returns

With the population of the world now well past the 3000 million mark, a simple division sum discloses the fact that each of us shares a birthday with some 9 million others (more than the population of Greater London!).

... that each of us shares a birthday with some nine million more others ...

A Welcome Guest

When the *Britannica*, the first steamer of the Cunard Line, arrived in Boston in 1890, her owner, Mr. CUNARD, received no less than two thousand invitations to dinner – enough to occupy his evenings continuously for the following five and a half years!

Which Watch?

Asked to judge between the reliability of a watch which lost ten minutes a day and one which had lost its main spring, you might reasonably decide in favour of the first. It is easy to calculate, however, that the losing watch would be correct only once in seventy-two days, whereas the other would show the right time twice every twenty-four hours!

Long Odds

The chance that a monkey, set before a typewriter and permitted to pound the keys at will, would eventually produce a typescript of the *Encyclopaedia Brittanica*, is certainly none too great. Nevertheless, it is with this kind of probability that modern information theory is principally concerned. The absurdity of the above situation is less likely to influence the theorist than the cold fact that the more information the *Encyclopaedia* contains, the less likely is the monkey to write the text.

Expensive Writing

Writing with a ball-point pen is as expensive as driving by car (per mile), neglecting the salaries of author and driver.

Smoking and High Living

Smoking ten cigarettes a day inactivates as much haemoglobin in the blood as does living at an altitude of 2000 metres (6500 feet) above sea level, due to the decreased air-pressure.

Hairsplitting

The wall of a soap bubble is some 100 atoms thick, but it is still about 10000 times thinner than the average human hair. Cynics might be tempted to observe that those who split atoms are therefore at least some million times as pedantic as those who split hairs!

Martial Music

The 35 million amateur musicians in the U.S.A. in 1964 could form a fourteen-piece band for each soldier on the American Army register.

Modern Migration

By the year 2000, a single, fine, summer weekend will see more people forsake their homes for the open (?) road in the Western European area bounded by London, Paris and Frankfurt, than were involved in the great movements of population within Europe throughout the 4th and 5th Centuries A.D.

This Odd Book – An Odd Time Scale

When the period of time which is thought to have elapsed since the creation of the cosmos (6×10^9 years) is divided into 100000 units, each of these *cosmic units of time* comprises 60000 years.

As it happens, the present ODD BOOK OF DATA can also be conveniently divided into approximately 100000 units, namely its composite letters, so that each of these may be given the value of 60000 years.

The second half of the text may then be taken to represent the period of time currently estimated to have passed since the Earth came into existence (roughly 3×10^9 years), whilst the last seventeen letters cover the million odd years of man's evolution to date.

The 60000 years represented by the final letter *s* of this article brought the cultural developments which made it possible to produce this book. If we wished to give these developments a more exact chronology, we should of course be obliged to cut the *s* into pieces, since this last letter of the text has a value

– 20 times greater than the age of our alphabet (3000 years);

– 120 times greater than the period elapsed since the invention of the printing press (1450);

– 200 times greater than the number of years during which the Modern English of the text has been current (three centurie*s*).

Basic Numerical Data

TABLE 1
Basic physical data

Symbol	Name	Value		
c	Velocity of light	2.997929	$\times 10^{10}$	cm sec^{-1}
e	Base of natural logarithms	2.7182818		
e	Electronic charge	4.80290	$\times 10^{-10}$	esu
h	Planck's constant	6.6253	$\times 10^{-27}$	erg sec
m	Electron rest mass[a]	9.1085	$\times 10^{-28}$	g
$m_n = M_n/N$	Neutron rest mass[b]	1.67473	$\times 10^{-24}$	g
$m_p = M_p/N$	Proton rest mass[c]	1.67242	$\times 10^{-24}$	g
N	Avogadro's number (phys. scale)	6.0248	$\times 10^{23}$	mole^{-1}
v	Velocity			
	– of sound	0.340		km sec^{-1}
	– Brownian motion (mean)	2		km sec^{-1}
	– electron in orbit	2.2	$\times 10^3$	km sec^{-1}
	Diameter H atom (in H$_2$)	0.75	$\times 10^{-8}$	cm
	Volume water molecule (in water)	3	$\times 10^{-23}$	cm^3

[a] 0.510976 MeV   [b] 939.51 MeV   [c] 938.21 MeV

TABLE 2
Metric equivalents

Name	Value		Name	Value	
Angström (Å)	10^{-8}	cm	Gallon	4.55	l
Inch	2.54	cm	BRT (1000 ft^3)	2.83	m^3
Foot	30.5	cm	Ton (warship)	1	m^3
Yard	0.914	m	Ounce	28.4	g
Mile	1.67	km	Lb	0.45	kg

TABLE 3
Astronomic data

Name	Value
Light second	3×10^5 km (1.86×10^5 miles)
Light minute	18×10^6 km (11.2×10^6 miles)
Light year	9.46×10^{12} km (5.9×10^{12} miles)
Parsec	3.26 light years
Distance between earth and nearest fixed star	$4\frac{1}{3}$ light years
Mean distance between earth and sun	$\sim 150 \times 10^6$ km (93×10^6 miles) $=$ $8\frac{1}{3}$ light min $= 108$ diam. of sun
Mean distance between earth and moon	$\sim 3.84 \times 10^5$ km (2.38×10^5 miles) $=$ $1\frac{1}{4}$ light sec $= 30$ diam. of earth
Mean diam. of earth	12740 km (7912 miles)
Diam. of sun	1.4×10^6 km (0.869×10^6 miles)
Diam. of moon	3476 km (2159 miles)
Rotation period of earth	23 h 56 min 40 sec
Siderial period of earth	365 d 5 h 48 min 46 sec
Mean speed of earth revolving about sun	29.76 km/sec (18.48 miles/sec)
Volume of earth	$\sim 10^{12}$ km^3 (2.4×10^{11} miles^3)
Surface of earth	510.1×10^6 km^2 (197×10^6 miles^2) (29% land, 71% water)
Mass of earth	5.97×10^{27} g
Mass density of earth	5.52 g/cm^3
Mass of sun	$\sim 332000 \times$ mass of earth
Mass of moon	$\sim 1/80 \times$ mass of earth
Gravitation constant	6.67×10^{-8} dyne cm^2 g^{-2}
Acceleration due to gravity (Brussels) (g)	981.13 cm/sec^2

TABLE 4

Time units

	Seconds	Minutes	Hours
Hour	3 600	60	1
Day	86 400	1 440	24
Year	31 536 000	525 600	8 760

TABLE 6

Energy conversion factors

	erg	joule = watt-sec	kilowatt-hour	cal
1 erg	1	$1 \cdot 10^7$	$2.78 \cdot 10^{-14}$	$2.39 \cdot 10^{-8}$
1 joule = watt-sec	$1 \cdot 10^7$	1	$2.78 \cdot 10^{-7}$	$2.39 \cdot 10^{-1}$
1 kilowatt-hour	$3.6 \cdot 10^{13}$	$3.6 \cdot 10^6$	1	$8.6 \cdot 10^5$
1 cal	$4.19 \cdot 10^7$	4.19	$1.16 \cdot 10^{-6}$	1
1 kgm	$9.8 \cdot 10^7$	9.8	$2.72 \cdot 10^{-6}$	2.34
1 electronvolt	$1.60 \cdot 10^{-12}$	$1.60 \cdot 10^{-19}$	$4.45 \cdot 10^{-26}$	$3.83 \cdot 10^{-1}$
1 mol-electronvolt	$9.65 \cdot 10^{11}$	$9.65 \cdot 10^4$	$2.68 \cdot 10^{-2}$	$2.3060 \cdot 10^4$
10^{-3} unit atomic mass	$1.49 \cdot 10^{-6}$	$1.49 \cdot 10^{-13}$	$4.14 \cdot 10^{-20}$	$3.56 \cdot 10^{-1}$
1 g-mass equivalent	$8.99 \cdot 10^{20}$	$8.99 \cdot 10^{13}$	$2.50 \cdot 10^7$	$2.15 \cdot 10^{13}$
1 megawatt-day	$8.64 \cdot 10^{17}$	$8.64 \cdot 10^{10}$	$2.4 \cdot 10^4$	$2.06 \cdot 10^{10}$

TABLE 5
Conversion factors

Name	Value	
hp h	270000	kgm
ft lb	0.1383	kgm
BTU (British Thermal Unit)	107.5	kgm
hp	75	kgm sec^{-1}
kg standard coal	7000	kcal

m	electron-volt	mol-electronvolt	10^{-3} unit atomic mass	g-mass equivalent	megawatt-day
$02 \cdot 10^{-8}$	$6.24 \cdot 10^{11}$	$1.04 \cdot 10^{-12}$	$6.70 \cdot 10^{5}$	$1.113 \cdot 10^{-2}$	$1.16 \cdot 10^{-18}$
$02 \cdot 10^{-1}$	$6.24 \cdot 10^{18}$	$1.04 \cdot 10^{-5}$	$6.70 \cdot 10^{12}$	$1.113 \cdot 10^{-14}$	$1.16 \cdot 10^{-11}$
$76 \cdot 10^{5}$	$2.25 \cdot 10^{25}$	37.3	$2.41 \cdot 10^{19}$	$4.01 \cdot 10^{-8}$	$4.17 \cdot 10^{-5}$
$27 \cdot 10^{-1}$	$2.61 \cdot 10^{19}$	$4.34 \cdot 10^{-5}$	$2.81 \cdot 10^{13}$	$4.66 \cdot 10^{-14}$	$4.84 \cdot 10^{-11}$
	$6.12 \cdot 10^{19}$	$1.02 \cdot 10^{-4}$	$6.57 \cdot 10^{13}$	$1.092 \cdot 10^{-13}$	$1.13 \cdot 10^{-10}$
$63 \cdot 10^{-20}$	1	$1.66 \cdot 10^{-24}$	$1.074 \cdot 10^{-6}$	$1.78 \cdot 10^{-33}$	$1.85 \cdot 10^{-30}$
$84 \cdot 10^{3}$	$6.03 \cdot 10^{23}$	1	$6.46 \cdot 10^{17}$	$1.074 \cdot 10^{-9}$	$1.11 \cdot 10^{-6}$
$52 \cdot 10^{-14}$	$9.31 \cdot 10^{5}$	$1.55 \cdot 10^{-18}$	1	$1.66 \cdot 10^{-27}$	$1.73 \cdot 10^{-24}$
$17 \cdot 10^{12}$	$5.61 \cdot 10^{32}$	$9.31 \cdot 10^{8}$	$6.03 \cdot 10^{26}$	1	$1.04 \cdot 10^{3}$
$81 \cdot 10^{9}$	$5.39 \cdot 10^{29}$	$8.99 \cdot 10^{5}$	$5.79 \cdot 10^{23}$	$9.61 \cdot 10^{-4}$	1